Routledge Introductions to Environment

Coastal Systems

Simon K. Haslett

London and New York

For Sam, Maya, Elinor and Rhiannon

First published 2000
by Routledge
11 New Fetter Lane, London EC4P 4EE

Simultaneously published in the USA and Canada
by Routledge
29 West 35th Street, New York, NY 10001

Reprinted 2003

Routledge is an imprint of the Taylor & Francis Group

Typeset in Times by Florence Production Ltd, Stoodleigh, Devon
Printed and bound in Great Britain by TJ International Ltd, Padstow, Cornwall

British Library Cataloguing in Publication Data
A catalogue record for this book is available from the British Library

Library of Congress Cataloging in Publication Data
Haslett, Simon K.
     Coastal systems / Simon K. Haslett.
         p. cm. – (Routledge introductions to environment series)
     Includes bibliographical references (p.).
     1. Coasts.   2. Coast changes.   3. Coastal zone management.   I. Title.
   II. Series.
   GB451.2.H38 2000
   551.45′7 – dc21   00–042464

ISBN 0–415–21301–0 (hbk)
ISBN 0–415–21302–9 (pbk)

Contents

Series editors' preface
Environmental Science titles

The last few years have witnessed tremendous changes in the syllabi of environmentally-related courses at Advanced Level and in tertiary education. Moreover, there have been major alterations in the way degree and diploma courses are organised in colleges and universities. Syllabus changes reflect the increasing interest in environmental issues, their significance in a political context and their increasing relevance in everyday life. Consequently, the 'environment' has become a focus not only in courses traditionally concerned with geography, environmental science and ecology but also in agriculture, economics, politics, law, sociology, chemistry, physics, biology and philosophy. Simultaneously, changes in course organisation have occurred in order to facilitate both generalisation and specialisation; increasing flexibility within and between institutions is encouraging diversification and especially the facilitation of teaching via modularisation. The latter involves the compartmentalisation of information which is presented in short, concentrated courses that, on the one hand are self-contained but which, on the other hand, are related to prerequisite parallel, and/or advanced modules.

These innovations in curricula and their organisation have caused teachers, academics and publishers to reappraise the style and content of published works. Whilst many traditionally styled texts dealing with a well defined discipline, e.g. physical geography or ecology, remain apposite, there is a mounting demand for short, concise and specifically focused texts suitable for modular degree/diploma courses. In order to accommodate these needs Routledge have devised the Environment Series which comprises Environmental Science and Environmental Studies. The former broadly encompasses subject matter which pertains to the nature and operation of the environment and the latter concerns the human dimension as a dominant force within, and a recipient of, environmental processes and change. Although this distinction is made, it is purely arbitrary and is made for practical rather than theoretical purposes; it does not deny the holistic nature of the environment and its all-pervading significance. Indeed, every effort has been made by authors to refer to such inter-relationships and to provide information to expedite further study.

This series is intended to fire the enthusiasm of students and their teachers/lecturers. Each text is well illustrated and numerous case studies are provided to underpin general theory. Further reading is also furnished to assist those who wish to reinforce and extend their studies. The authors, editors and publishers have made every effort to provide a series of exciting and innovative texts that will not only offer invaluable learning resources and supply a teaching manual but also act as a source of inspiration.

A. M. Manion and Rita Gardner
1997

Series International Advisory Board

Australasia: Dr P. Curson and Dr P. Mitchell, Macquarie University

North America: Professor L. Lewis, Clark University; Professor L. Rubinoff, Trent University

Europe: Professor P. Glasbergen, University of Utrecht; Professor van Dam-Mieras, Open University, The Netherlands

● Plates

Figures

Tables

Boxes

Author's preface

In 1998 I undertook fieldwork along the coast of northern Queensland in Australia, and at each beach I visited there were signs alerting visitors to various hazards. These included dangerous currents, stinging jellyfish, saltwater crocodiles, sunburn, and even falling coconuts! Being forewarned, I was able to enjoy investigating the various locations. However, there is a very serious message embedded in this, that coasts command respect. In recent years there have been a number of major coastal disasters resulting in thousands of casualties, including the Papua New Guinea tsunami in 1998, the Orissa storm surge in the Bay of Bengal in 1999 and the devastating flooding of the coastal lowlands of Mozambique in 2000 (some of these are discussed further in this book).

Around 3 billion people live in the coastal zone, that is half of the world's entire population. It is no wonder then that coasts are under pressure. Indeed, it is a major challenge for the twenty-first century for humans to live and work at, and exploit, coasts in a way that does not damage the environment or deplete resources, that we will pass on to our children that which we inherited from previous generations. If we are to achieve this goal of sustainable development, then we should be striving to understand better how physical and ecological coastal systems operate, and then work alongside these, and not battle against them.

It has been my intention in writing this book to introduce to the reader our current understanding of coastal systems, including the physical, ecological and human interactions, so as to raise an awareness of the diversity and sensitivity of these precious environments. To conform with the 'Introductions to Environment' series this book has been written for first and second year undergraduate audiences studying coastal systems as a one-semester module. Out of necessity therefore, the text proceeds from first principles, but I hope that there is sufficient detail herein to satisfy more advanced readers. To this end, I have carefully selected references for further reading.

I have been supported for many years in my oceanographic research by a large number of people and to all of them I am very grateful. In particular, I am indebted to my wife Sam and my children Maya, Elinor and Rhiannon, for their patience, encouragement and companionship in my oceanic endeavours, and to my parents for fostering my environmental interests from a very early age. My professional development has benefited greatly from my mentors, Brian Funnell, John Murray, Rick Curr, Ted Bryant, Annika Sanfilippo and Allan Williams. All my colleagues at Bath Spa University College have offered very sympathetic support to me whilst I have been writing this book, and especially Paul Davies and Fiona Strawbridge who have been brave enough to embark on coastal research collaborations with me. I am also very grateful to Kevin Kennington, Andy Cundy and Chris Spencer for taking the time to digest the first draft of the text and to offer very constructive comments. Finally, thanks to Ann Michael and Casey Mein at Routledge for making the writing of this book as painless as possible.

Simon Haslett
Newton Park, Bath
20 April 2000

Note on the text

Bold is used in the text to denote words defined in the Glossary. It is also used to denote key terms.

Acknowledgements

The author and publisher would like to thank the following for granting permission to reproduce images in this work:

David Smith and Alastair Dawson for the line drawing of tsunami wave generation, reprinted from 'Tsunami waves in the North Sea', *New Scientist*, 4 August 1990, with permission from *New Scientist*.

R. A McBride *et al.* for Figure 2, reprinted from *Marine Geology*, Volume 126, McBride *et al.*, 'Geomorphic response-type model for barrier coastlines: a regional perspective', pp. 143–159, 1995, with permission from Elsevier Science.

G. Masselink and C. B. Pattiaratchi for Figure 1, reprinted from *Marine Geology*, Volume 146, Masselink *et al.*, 'Morphological evolution of beach cusps and associated swash circulation patterns', pp. 93–113, 1998, with permission from Elsevier Science.

J. R. L. Allen for Figures 2 and 5, reprinted from *Proceedings of the Geologists' Association*, Volume 104, J. R. L. Allen, 'Muddy alluvial coasts of Britain', pp. 241–262, 1993 and Figures 1 and 2, reprinted from *Proceedings of the Geologists' Association*, Volume 107, J. R. L. Allen, 'Shoreline movement and vertical textural patterns . . .', pp. 15–23, 1996, both published with permission of the Geologists' Association.

H. G. Reading, J. D. Collinson, Blackwell Science and The Houston Geological Society for Figure 6.2 and H. G. Reading, J. D. Collinson, Blackwell Science for Table 6.2, reprinted from the chapter 'Clastic Coasts' in the book *Sedimentary Environments*, 3rd edition, edited by H. G. Reading, 1996.

R. W. Young *et al.* and *Australian Geographer* for Figure 1, reprinted from *Australian Geographer*, Volume 27, R. W. Young *et al.*, 'Fluvial deposition on the Shoalhaven Deltaic Plain, southern New South Wales', pp. 215–234, 1996.

R. A. Warrick *et al.* and the Intergovernmental Panel on Climate Change for Figure 7.8 and Table 7.8, from the chapter 'Changes in sea level', published in *Climate Change*, edited by J. T. Houghton *et al.*, 1996.

J. M. Hooke, M. J. Bray and the Royal Geographical Society for Figure 1, reprinted from *Area*, Volume 27(4), J. M. Hooke and M. J. Bray, 'Coastal groups, littoral cells, policies and plans in the UK', p. 362, 1995.

The Countryside Agency for permission to extract data from their Heritage Coasts Factfile web site at http://www.countryside.gov.uk/what/hcoast/heri_tbl.htm

D. Briggs *et al.* and Routledge for Figures 1 (in Box p. 308), 1.6, 1.10, 3.5, 4.7, 4.11, 4.12, 4.13, 12.12, 16.1, 16.6, 17.1, 17.2b, 17.3, 17.4, 17.5, 17.9, 17.12, 17.13, 17.14, 17.16, 17.17, 17.18 and 17.20 reprinted from *Fundamentals of the Physical Environment*, 2nd edition, 1997. Further permission to reproduce Figures 16.6b and 17.5 has been obtained from Academic Press, the publishers of R. W. G. Carter, *Coastal Environments*, 1988; and Blackwell Publishers for Figure 17.12, the publishers of A. Goudie, *The Nature of the Environment*, 1984; and Arnold for Figures 4.12 and 16.1, the publishers of C. A. M. King, *Oceanography for Geographers*, 1962, and A. Warren, Aeolian Processes in C. Embleton and J. Thornes (eds) *Processes in Geomorphology*, 1979; and the Institute of Terrestrial Ecology for Figure 17.20, who are the copyright holders of L. A. Boorman *et al.*, *Climate change, rising sea level and the British coast*, 1989; and the Houston Geological Society for Figure 17.9, the publishers of Galloway (1975); and Pearson Education Ltd for Figures 17.16, 17.17, 17.18, who are the copyright holders of J. L. Davies, *Geographical Variation in Coastline Development*, 1980; and Taylor & Francis for Figure 17.5, who are the copyright holders of R. W. G. Carter, *Boreas*, volume 12, pages 167–182, 1983

P. French and Routledge for Figures 2.3, 2.8, 2.10 and 2.11, reprinted from *Coastal and Estuarine Management*, 1997.

R. Kay, J. Alder and E & FN Spon for figures from Boxes 4.4 and 4.16, reprinted from *Coastal Planning and Management*, 1999.

C. Park and Routledge for Figures 9.10, 11.25, 13.8, 15.4, 15.5, 15.12, 15.13, 15.14 and Table 11.9, reprinted from *The Environment*, 1997. Further permission to reproduce Figure 9.10 has been obtained from Academic Press, the publishers of R. W. G. Carter, *Coastal Environments*, 1988; and McGraw-Hill for Figures 11.25, 13.8, 15.5, 15.13, 15.14, the copyright holders of A. H. Doerr, *Fundamentals of Physical Geography*, 1990; and Longman for Figure 15.12, who are the publishers of C. Buckle, *Landforms in Africa*, 1978.

K. Pickering, L. Owen and Routledge for Figures 8.7 and 8.11b, reprinted from *An Introduction to Global Environmental Issues*, 2nd edition, 1997.

Introduction

The coast is a very special environment in that it is where the land, sea and atmosphere meet. Each of these contributes to the workings of the coast, making coasts very interesting, and yet a challenging subject to study. The coast is also the location of major human settlements, and human activity can have significant impacts on the operation of coasts to the point of environmental and socio-economic degradation. The study of coasts, therefore, is highly interdisciplinary, incorporating geology, physical and human geography, oceanography, climatology, sociology, economics, engineering, planning, management, and so on. It is perhaps one of the best examples of interdisciplinary environmental science.

It is clear that coasts are not isolated environments, they receive energy and material, process these inputs, and subsequently may lose them as output. Variations in inputs cause changes in the physical environment, for example, an increase in wave energy may enhance coastal erosion, or a decrease in nutrients may limit biological productivity, so demonstrating that all natural coastal operations are interlinked in some way. This describes a system, and the study of coasts is well suited to a systems approach.

The underlying theme throughout this book is the operation of coastal systems, explicitly demonstrated at times, and implied at others. However, in the real world the workings of a system are largely hidden from view and must be interrogated through field evidence, such as landforms, animals and plants. The recognition of this evidence by a student is important, so that a significant part of this book describes and illustrates a range of coastal landforms and ecosystems. At present, however, one of the most exciting aspects of coastal studies is system dynamics, investigating the response of the physical environment to external forces, such as wave energy, sea level change and human interference. In exploring dynamics here, one will hopefully gain an intimate understanding of the inter-relationships within the coastal system, therefore, much of the text is devoted to this.

The diversity of coastal environments is a major attraction for their study. An introductory text like this, however, cannot hope to cover this variety, either in breadth or depth. Therefore, the book is divided based on the dominant processes operating on a given coastline, be it waves, tides or rivers. The role of sea level is also covered, and with its increased prominence in the media since the late 1980s, as a consequence of global warming, a relatively detailed explanation of how sea level curves are produced and their accuracy is given, as are the mechanisms of sea level rise under climate change and its management. Further discussion of management issues is the subject of the final chapter. However, case studies of human impact on the coast and associated management issues are scattered throughout the text where appropriate.

Finally, in recent years there has been a growing appreciation of the variation in the rate of coastal change. For many years it was held that coasts developed more or less exclusively through the gradual (high frequency/low magnitude) action of waves, tides and wind. However, many scientists now suggest that low frequency/high magnitude events, such as storms and tsunami, may play dominant roles in the evolution of some coasts. I have attempted to provide an unbiased approach, incorporating both of these opposing views.

1 Coastal systems: definitions, energy and classification

The space occupied by the coast is not easily defined. It is a complex environment that has attributes belonging to both terrestrial and marine environments, which defies a truly integrated classification. This chapter covers:

- the definition of the coast from scientific, planning and management standpoints
- the sources of energy that drive coastal processes
- the architecture and working of coastal systems, introducing concepts of equilibrium and feedbacks
- an introduction to coastal classifications, with an emphasis on broadscale geological and tectonic controls
- a discussion of the complexities of terminology used in studying coastal systems

Introduction

Defining the coast

The coast is simply where the land meets the sea. However, applying this statement in the real world is not that straightforward. It is not always easy, for instance, to define exactly where the land finishes and the sea begins. This is particularly so for extensive low-lying coastal wetlands, which for most of the time may be exposed and apparently terrestrial, but a number of times a year become submerged below high tides – does this environment belong to the sea or to the land, and where should the boundary between the two be drawn? It is much more meaningful therefore, not to talk of coastlines, but of **coastal zones**, a spatial zone between the sea and the land. Usefully, this has been defined as the area between the landward limit of marine influence and the seaward limit of terrestrial influence (Carter 1988). If we accept this definition, then coasts often become wide spatial areas, for example, encompassing land receiving sea-spray and blown sand from beach sources, and out to sea as far as river water penetrates, issued from estuaries and deltas.

Box 1.1

Definitions of the coastal zone for planning and management

The definition of the coastal zone given on p. 3 is very much for the use of physical scientists studying the coast. However, for planning and management purposes, where administration is involved, the coastal zone is much more variably defined. Kay and Alder (1999) give a range of definitions used by various organisations in international and national government. Some definitions are known as distance definitions, whether fixed or variable, where the coastal zone is defined as being so many kilometres landward, and so many nautical miles seaward, of the shoreline. Other definitions do attempt to recognise and incorporate aspects of the working complexity of the coastal zone. In abbreviated form, these include:

- 'the coastal waters and the adjacent shorelands strongly influenced by each other, and includes islands, transitional and intertidal areas, salt marshes, wetlands and beaches. The zone extends inland from the shorelines only to the extent necessary to control shorelands, the uses of which have a direct and significant impact on the coastal waters' (United States Federal Coastal Zone Management Act)
- 'as far inland and as far seaward as necessary to achieve the Coastal Policy objectives, with a primary focus on the land–sea interface' (Australian Common-wealth Coastal Policy)
- 'definitions may vary from area to area and from issue to issue, and that a pragmatic approach must therefore be taken' (United Kingdom Government Environment Committee)
- 'the special area, endowed with special characteristics, of which the boundaries are often determined by the special problems to be tackled' (World Bank Environment Department).

Coastal energy sources

Coasts are not static environments and are in fact highly dynamic, with erosion, sediment transport and deposition all contributing to the continuous physical change that characterises the coast. Such dynamism requires energy to drive the coastal processes that bring about physical change, and all coasts are the product of a combination of two main categories of processes driven by different energy sources (Figure 1.1).

1 The first category of processes is known as the **endogenetic processes** and are so called because their origin is from within the earth. Endogenetic processes are driven by geothermal energy which emanates from the earth's interior as a product of the general cooling of the earth from its originally hot state, and from radioactive material, which produces heat when it decays. The flux of geothermal energy from the earth's interior to the surface is responsible for driving continental drift and is the energy source in the plate tectonics theory. Its

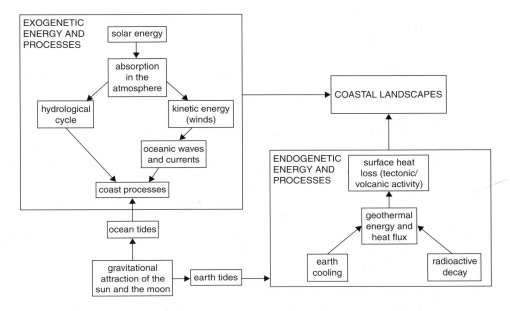

Figure 1.1 *Endogenetic and exogenetic energy and processes and their contribution to the development of coastal landscapes.*

influence on the earth's surface, and the coast is no exception, is to generally raise **relief**, that is to generally elevate the land.

2 The second category of processes is known as **exogenetic processes**, that is those processes that operate at the earth's surface. These processes are driven by solar energy. Solar radiation heats the earth's surface which creates wind, which in turn creates waves. It also drives the **hydrological cycle**, a major cycle in the evolution of all landscapes, and describes the transfer of water between natural stores, such as the ocean. It is in the transfer of this water that rain falls and rivers flow, producing important coastal environments, such as estuaries and deltas. The general effect of exogenetic processes is to erode the land, such as erosion by wind, waves and running water, and so these processes generally reduce relief (however, sand dunes are an exception to this rule, being built up by exogenetic processes).

A third source of energy that is important for coasts is that produced by the gravitational effects of the moon and sun. Principally such gravitational attraction creates the well-known ocean tides which work in association with exogenetic processes, but also they produce the lesser-known earth tides which operate in the molten interior of the earth which assist the endogenetic processes.

Ultimately, all coastal landscapes are the product of the interaction of these broadscale process categories, so where endogenetic processes dominate, mountainous coasts are often produced, whereas many coastal lowlands are dominated by exogenetic processes. Commonly, however, there is a more subtle balance between the two, with features attributable to both process categories present.

Coastal systems

Natural environments have for some time been viewed as systems with identifiable inputs and outputs of energy (a **closed system**) or both energy and material (an **open system**), and where all components within the system are inter-related (Briggs *et al.* 1997). The boundaries of a system are not always easily defined, as we discovered above when trying to define the coast. Where we can identify a relationship between inputs and outputs, but do not really know how the system works, then we are dealing with a **black box system** (Figure 1.2); the coast as a whole may be viewed as a black box system. A study of the system may reveal a number of subsystems within it, linked by flows of energy and matter, known as a **grey box system**; a coastal example of this may be a cliff system being eroded by wave energy, which then supplies an adjacent beach system with sediment. Further investigation may reveal the working components of the system, with energy and material pathways and storages, known as a **white box system**. Following on from our previous example, these components may include the rock type that the cliff is composed of, the type of erosion operating on the cliff, sediment transport from the cliff to the beach, beach deposition and its resulting morphology.

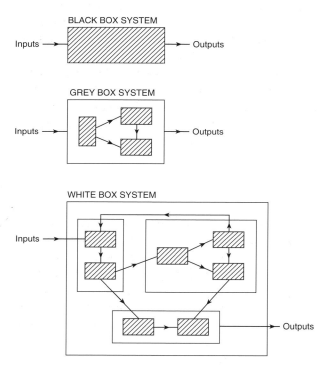

Figure 1.2 *Types of systems*.
Source: Briggs *et al.* (1997) (Figure 1.6, p. 5)

System approaches

At the finest scale, then, a system comprises components that are linked by energy and material flows. However, there are four different ways in which we can look at physical systems:

1 **Morphological systems** – this approach describes systems not in terms of the dynamic relationships between the components, but simply refers to the morphological expression of the relationships. For example, the slope angle of a coastal cliff may be related to rock type, rock structure, cliff height, and so on.

2 **Cascading systems** – this type of system explicitly refers to the flow or cascade of energy and matter. This is well exemplified by the movement of sediment through the coastal system, perhaps sourced from an eroding cliff, supplied to a beach, and then subsequently blown into coastal sand dunes.

3 **Process-response systems** – this combines both morphological and cascading systems approaches, stating that morphology is a product of the processes operating in the system. These processes are themselves driven by energy and matter, and this is perhaps the most meaningful way to deal with coastal systems. A good example is the retreat of coastal cliffs through erosion by waves. Very simply, if wave energy increases, erosion processes will often be more effective and the cliff retreats faster. It is very clear from this example that the operation of a process stimulates a morphological response.

4 **Ecosystems** – this approach refers to the interaction of plants and animals with the physical environment, and is very important in coastal studies. For example, grasses growing on sand dunes enhance the deposition of wind-blown sand, which in turn builds up the dunes, creating further favourable habitats for the dune biological community, and indeed may lead to habitat succession.

The concept of equilibrium

Coasts are dynamic and they change frequently. These changes are principally caused by changes in energy conditions, such as wave energy, for example, which may increase during storms. The morphology of the coast responds to changes in energy because it aims to exist in a state of equilibrium with the reigning processes (Briggs *et al.* 1997). However, there are three types of equilibrium (Figure 1.3):

1 **Steady state equilibrium** – this refers to a situation where variations in energy and the morphological response do not deviate too far from the long-term average. For example, along a coast that experiences relatively consistent wave energy conditions, the gradient of a beach may be steeper at certain times of the

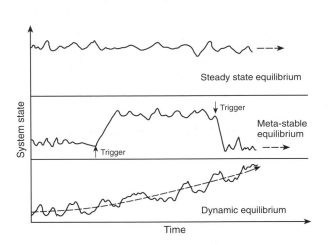

Figure 1.3 *Types of equilibrium.*
Source: Briggs *et al.* (1997) (Figure 1.10, p. 7)

year, and shallower at others, but the average annual gradient is similar from year to year.

2 **Meta-stable equilibrium** – this exists where an environment switches between two or more states of equilibrium, with the switch stimulated by some sort of trigger. An example of this includes the action of high energy events, such as storms or tsunami, which can very rapidly switch a coastal system from one state of equilibrium to another, by removing or supplying large volumes of beach sediment for example. Also, human activity often has this effect on coastal environments.

3 **Dynamic equilibrium** – like meta-stable equilibrium, this too involves a change in equilibrium conditions, but in a much more gradual manner. A good example is the response of coasts to the gradual rise in sea levels that were experienced through the twentieth century as a result of climate change.

System feedbacks

Understanding states of equilibrium in a system requires some knowledge of feedbacks within a system. Feedbacks occur as the result of change in a system and they may be either positive or negative, respectively switching the system to a new state of equilibrium or attempting to recover to the system's original state of equilibrium. **Positive feedbacks** therefore, tend to amplify the initial change in the system, so that, for example, the ridge of a coastal sand dune breached by storm wave erosion may be subsequently laterally undercut by wind erosion, thus fragmenting the dune ridge and leaving it more susceptible to further wave erosion (Figure 1.4a). Ultimately, the entire dune ridge may be relocated further inland and a new state of equilibrium reached. **Negative feedbacks**, however, tend to dampen the effect of the change. For example, sand eroded during a storm from the front of sand dunes at the back of a beach, may be redeposited as offshore sand bars which help to protect the beach-dune system from further storm waves, by reducing the amount of wave energy reaching the dune front (Figure 1.4b). Managing coastal systems requires a detailed knowledge of feedbacks, as all too frequently, as we shall see throughout this book, human intervention in one part of a coastal system often leads to a number of apparently unforeseen and undesirable feedbacks.

The classification of coasts

Because there is such a wide variety of factors that affect coasts, it has been very difficult to actually create an integrated classification scheme. As a result there has been a number of attempts to classify coasts according to single parameters, such as wave or tidal environment, geology, and tectonic setting. Waves and tides are covered individually in other chapters in this book (see Figures 2.1, 3.4 and 3.6),

(a)

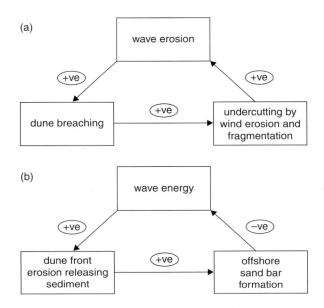

(b)

Figure 1.4 *Examples of feedback relationships in coastal systems: (a) example of positive feedback where a positive relationship exists between wave erosion and dune breaching (meaning that an increase/decrease in one, i.e. wave erosion, increases/decreases the other, i.e. the amount of dune breaching), which in turn has a positive relationship to subsequent dune undercutting through wind erosion, so further fragmenting the dune ridge and leaving it more susceptible to further wave erosion. Positive feedback involves either no negative relationships, as in this example, or an even number of them; (b) example of negative feedback. Note that a negative relationship, such as here between offshore sand bar formation and wave energy, means that as one increases/decreases the other decreases/increases in response. Negative feedbacks always involve an odd number of negative relationships.*

and furthermore are usually only applicable on the local to regional scale. Here we will concentrate on the broadscale geological and tectonic settings of coasts, which often are applicable to coasts along entire continental margins.

Geological classification

In 1888 E. Suess put forward a coastal classification based on geological structure and its orientation as regards the general trend of the coastline. On this premise he recognised two types of coasts:

1 **Pacific type** – where the orientation of the rock structure lies parallel to the coastline. This type of coast is also known as accordant or concordant, and often forms rather straight coastlines interupted by relatively small embayments. The Dalmatian coast of the former Yugoslavia in the Mediterranean, is an excellent example of a Pacific type coast.

2 **Atlantic type** – where the orientation of the rock structure is at right angles (perpendicular) to the coastline. This coastal type is often known as discordant, and is characterised by prominent headlands and embayments. The southwest coast of Ireland, including Bantry Bay, is a good example of this coastal type.

Horsfall (1993) explores the application of this classification to the classic coastal scenery of Dorset, England. Although it is useful, this scheme is limited in that it gives no indication as to the dynamics of a coastline, is it submerging/emerging, or is deposition/erosion dominant? It is just a statement of the relationship between rock structure and general coastline orientation, which is a random relationship. Small-scale structure can also influence the character of a coast, such as the attitude of joints in granite (Plate 1.1).

Plate 1.1 *Small-scale rock structure influences the character of a coastline. Here, along the granite coast of western Brittany (France), two contrasting coasts exist: (a) at Concarneau the intertidal zone has a very smooth appearance because the joints in the granite are quasi-horizontal, whereas (b) at Pointe Karreg Leon, near Audierne, the intertidal zone is very jagged because the jointing is quasi-vertical.*

Box 1.2

Integrating the geological classification

Bishop and Cowell (1997) have investigated the relationship between rock structure, river drainage patterns, sea level change, and sedimentology along the eastern Australian coastline. They confirm that Pacific type coasts tend to possess abundant small embayments, whereas embayments are larger on other coasts. However, their analysis suggests that in general, embayment size is related to the size of the river draining into the embayment rather than geological control. They also suggest that sea level has an influence on coastal morphology here, with higher sea levels resulting in a more crenulate and compartmentalised coast with smaller embayments, and lower sea levels creating longer embayments and a more open, less compartmentalised coast. This study does much to try and integrate geology into a more meaningful understanding of coastal development.

Tectonic classification

The theory of plate tectonics describes the creation and destruction of crustal material, and in doing so it explains the movement of the continents around the globe. The earth's surface comprises a number of continental and oceanic crustal plates (Figure 1.5). Each one of these plates is bounded by zones where either new crust is created (constructive boundary) or where old crust is destroyed (destructive boundary). Plate movement is usually away from constructive boundaries (divergence), and towards destructive boundaries (convergence). At the broadscale, there is often a consistent relationship between the characteristics of a coastline and the type of plate boundary that it is nearest to. Inman and Nordstrom (1971) recognise a number coastal types based on the plate boundary they are associated with. Davis (1994) reviews **leading edge coasts**, **trailing edge coasts**, and **marginal sea coasts**, all of which are briefly described below.

Leading edge coasts

These occur where a continental plate converges with an oceanic plate at a destructive boundary. Because of this, these types of coasts are also known as convergent margin coasts. Continental crust is less dense and, therefore, more buoyant than oceanic crust resulting in the oceanic plate going under or subducting beneath the continental plate. The compressional forces created by convergence cause the rocks along the coast to buckle, fold and fault, uplifting them to create chains of coastal mountains. Earthquakes are commonly associated with this coastal type (see Box 1.3). The continental shelf in front of the mountain chain is usually

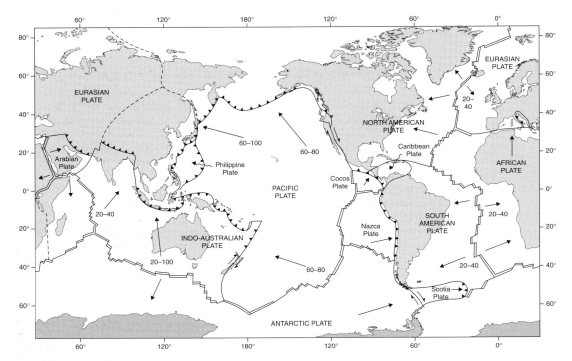

Figure 1.5 *The distribution of the earth's crustal plates and their various boundary types. Parallel lines = constructive (divergent) boundary, toothed lines = destructive (convergent) boundary (teeth pointing towards subduction), and single/broken lines = conservative (transform) boundary. Arrows and figures indicate plate movement in mm y⁻¹.*
Source: Briggs *et al.* (1997) (Figure 3.5, p. 36)

narrow or even absent, and therefore, the gradient from the top of the coastal mountains to the sea floor is usually very steep. Therefore, although much sediment is eroded from the uplifting coastal mountains by fast-running streams, it is usually lost into deep water via submarine canyons when it is introduced into the sea. Commonly, then, leading edge coasts are characteristically mountainous, dominated by erosional processes, and so often rocky. The longest leading edge coast occurs more or less along the entire western seaboard of the Americas (Plate 1.2), both northern and southern continents, with the exception of southern California.

Trailing edge coasts

These are coasts that form as plates rift apart due to divergence, allowing the ocean to inundate the rift to create a new sea. Once formed, these coasts are carried away from the diverging boundary as passive continental margins (Plate 1.3). Shortly after rifting the coasts comprise relatively high relief and possess a fairly steep

Box 1.3

Tectonic compression, earthquakes and leading edge coasts

Leading edge coasts are subject to compressional tectonic forces produced as the two plates converge. Plates do not move past each other continuously, but through sudden movements – seismic events, such as an earthquake. Earthquakes represent the release of the energy built up by the frictional strain of the two plates. During earthquakes the coast often sinks, sometimes by over 1 m, a phenomenon known as **coseismic subsidence**. In the period between earthquakes compressional energy builds up again, referred to as **interseismic strain accumulation**, which has the effect of raising the land back out of the sea. Therefore, leading edge coasts often go through a cycle of rapid submergence during an earthquake, followed by a period of gradual elevation. This process is often dramatically seen with the drowning of coastal forests that colonised the coast during the interseismic period, which may last several hundreds of years. Long and Shennan (1998) provide an example from the Washington–Oregon coast, USA.

Plate 1.2 *The rugged coast of Oregon (USA), looking north towards Hecata Head lighthouse on Highway 101, is a good example of a leading edge coast being dominated by erosion with very limited coastal deposits.*

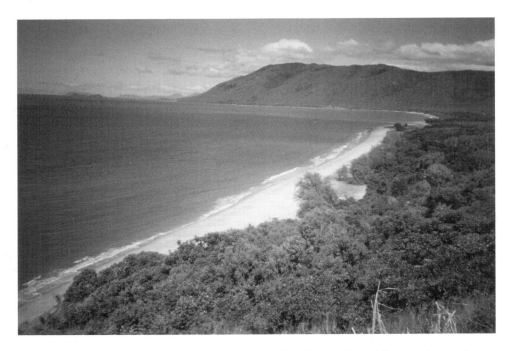

Plate 1.3 *The trailing edge coast of eastern Australia (this section of the coast is to the north of Cairns, Queensland). Although at first glance it is similar to a leading edge coast, the continental shelf is wider here, supporting more extensive coastal deposits and, indeed, the Great Barrier Reef situated offshore.*

gradient, and in many respects their topography is very similar to leading edge coasts. These are known as **neo-trailing edge coasts** and the present-day coasts of the Red Sea belong to this subdivision. As divergence progresses, the sea expands and erosion of the coast increases, both by wave activity at the shoreline, and through the action of high energy streams flowing down steep hills. In this way, the continental shelf begins to widen, the relief is lowered, and the **afro-trailing edge coast** is created. As the name implies, most of the African coastline is of this type, with the exception of course of some Mediterranean and Red Sea coasts. Africa has been tectonically stable for many millions of years, and although the continental shelf is relatively wide now, large sedimentary features such as deltas are limited in comparison with the most mature trailing edge coastal subtype, the **amero-trailing edge coast**. This mature coastal type is characteristic of the eastern seaboard of the Americas, with an extensive coastal plain existing along the North American section, and vast deltas, such as the Amazon Delta, characterising the South American section.

Marginal sea coasts

This is a relatively uncommon coastal type and occurs where plate convergence takes place offshore, with a relatively wide continental shelf separating the plate boundary from the coast. In many ways, this coastal type is similar to trailing edge coasts, especially in the sedimentary features that may develop, but it suffers earthquakes more regularly and may experience limited tectonic movement. The Gulf of Mexico coasts are an example, with the volcanic islands of the Caribbean marking the zone of plate convergence.

Coastal terminology

As with all scientific disciplines, the study of coasts has generated a vast lexicon of terms to describe landforms, processes, deposits, habitats, and ecosystems. Hopefully readers will agree that one of the most valuable parts of this book is the fairly extensive glossary. I have included all the important coastal terms that one will come across in the literature, in the hope that students will not be discouraged by the sometimes bewildering array of terms that are used, not only to describe different attributes of coastal systems, but sometimes apparently the same attributes. Rarely, however, are these terms precisely synonymous, and important, if subtle, differences in meaning do exist. A command of coastal terminology serves to better understand coastal systems, and also to enhance the communication of the subject. I therefore encourage you not to pass over unfamiliar terms, but to try and grasp their meaning both from the text in which it occurs, and with reference to the glossary. For when one goes beyond this introductory text to delve into the primary literature in scientific journals, which I thoroughly recommend, one quickly discovers that no allowance is given to readers with a limited vocabulary on the subject, and the value of spending time reading such articles is correspondingly diminished.

A useful example of similar yet subtly different terms involves the standard subdivision of the coastal zone (Figure 1.6). Two sets of terms are shown. However, one subdivides the coastal zone based on morphological changes (**backshore, foreshore, inshore**, and **offshore**), whilst the other is based on the types of wave processes that operate in different parts of the coastal zone (**swash zone, surf zone**, and **breaker zone**, which together comprise the **nearshore zone**). The same space may also be subdivided on sedimentological grounds, either according to the type of sediment, sedimentary structures, and/or depositional processes that occur (Reading and Collinson 1996):

- the **beach zone** occurs between mean low water and the landward limit of wave activity and mainly possesses parallel bedded sand layers.
- the **shoreface zone** occurs between mean low water and the mean fairweather

Figure 1.6 *Subdivision of the coastal zone using both morphological and wave process terminology.*
Source: Briggs *et al.* (1997) (Figure 17.1)

wave base (see pp. 20 and 50 for explanation) and is characterised by wave-swept sand ripples.

- the **offshore-transition zone** lies between the mean fairweather wave base and the mean storm wave base (again, see pp. 20 and 50) and is dominated by sediment deposition during storms.
- the **offshore zone** lies below the mean storm wave base and is characterised by fine-grained sediment settling out of the water.

Summary

- The coastal zone may be defined as the area between the landward limit of marine influence and the seaward limit of terrestrial influence.

- Coasts, like all landscapes, are the product of the combined influence of endogenetic and exogenetic energy and processes.

- Coasts operate as open systems with the flow of energy and material into, through and out of the coastal system. Coastal morphology aims to attain a state of equilibrium with the reigning processes. However, change within the system may trigger feedbacks, which may either accentuate the change (positive feedback) or help to minimise its effect (negative feedback).

- The classification of coasts is problematic, but on the broadscale the use of tectonic conditions helps us subdivide the global coastline into leading edge, trailing edge, and marginal sea coasts, each possessing a range of characteristic attributes.

- The use of diverse and sometimes confusing terminology in studying coasts is important for the student to command.

Discussion questions

1 Examine the problems in constructing a meaningful definition of the coast.

2 The study of coasts is well suited to a systems approach. Discuss why this might be so.

3 Assess the significance of tectonics in the development of a global coastal classification scheme.

Further reading

See also

Administrative coastal management issues, Chapter 6
The development of shore platforms, Chapter 2
Wave classification of coasts, Chapter 2
Tidal classifications of coasts, Chapter 3
Wave and beach terminology, Chapter 2

General further reading

Coasts: An Introduction to Coastal Geomorphology (3rd Edition). E. C. F. Bird. 1984. Blackwell, Oxford, 320pp. A standard introduction to coastlines.

Coastal Environments: An Introduction to the Physical, Ecological and Cultural Systems of Coastlines. R. W. G. Carter. 1988. Academic Press, London, 617pp. A thorough advanced-level overview of the coast, quite technical in places, and becoming out-of-date.

The Evolving Coast. R. A. Davis, Jr. 1994. Scientific American Library, New York, 231pp. A concise introduction to the physical development of coastlines.

Coasts. J. D. Hansom. 1988. Cambridge University Press, Cambridge, 96pp. A concise and readable introduction to coasts.

Les littoraux: Impact des aménagements sur leur évolution. R. Paskoff. 1998. Armand Collin, Paris, 260pp. An excellent French language text dealing with coastal problems.

Introduction to Coastal Geomorphology. J. Pethick. 1984. Arnold, London, 260pp. Although titled 'Introduction' this is in fact a fairly technical, and now slightly out-dated, guide to coastal geomorphology, but highly recommended for Pethick's elegant writing and excellent diagrams.

Exploring Ocean Science (2nd Edition). K. Stowe. 1996. Wiley, New York, 426pp. A very well illustrated and thorough review of oceanography. Only some of the chapters deal with the coastal ocean, but covers processes, landforms and ecology.

Coastal Dynamics and Landforms. A. S. Trenhaile. 1997. Oxford University Press, Oxford, 365pp. An advanced-level text dealing with process and form at the coast.

The periodicals *Journal of Coastal Research*, *Estuarine, Coastal and Shelf Science*, *Shore and Beach*, *Journal of Coastal Conservation*, *Marine Geology* and others should be consulted regularly for individual research papers, themed sections, and special issues devoted to particular topics.

2 ▶ Wave-dominated coastal systems

All coasts are affected to a certain degree by wave activity, and waves provide the energy that drives many of the coastal processes that create many of the world's most spectacular coasts. This chapter covers:

- **the origin and characteristics of waves and their processes**
- **the dynamic interaction between waves and coastal systems**
- **the geomorphology and ecology of erosional and depositional wave-dominated coastlines**
- **examples of environmental, management and engineering issues affecting wave-dominated coastlines**

Introduction

Wave-dominated coasts are amongst the most familiar of all coastal environments. Sandy beaches are popular holiday destinations and dramatic cliffed coastlines are often frequented by walkers, hikers, rock-climbers and hang-gliders. They are very dynamic and sensitive systems, often in equilibrium in the natural state, but susceptible to even slight interference from human activity.

On a global scale, coastal wave environments reflect to a large degree the climatic conditions experienced in a given region, such as wind speed and duration, as well as the size of the ocean concerned (Figure 2.1). For example, the mid-latitudes in both northern and southern hemispheres experience regular storms which in turn can produce very large waves. Storm wave affected coasts, therefore, occur in the path of major storm tracks, such as northwest North America from Oregon to Alaska, northeast North America from Florida to Newfoundland, northwest European coasts of France, British Isles and Scandinavia, Atlantic and Pacific coasts of southern South America and southerly coasts of Australia and New Zealand. Storm wave affected coasts also occur in tropical latitudes, such as southeast Asia and the Arabian Sea, where they regularly experience tropical cyclones. Outside these storm belts, waves on open ocean coasts tend to be smaller, but may still be significant in influencing coastal development. This is at a minimum, however, in the subtropical doldrums, where wind activity is least intense. Enclosed sea coasts, such as the Mediterranean and Red Seas, experience minimum wave conditions due to their size, which prohibits the development of large waves (see next section). However, it is coasts in polar regions that are least affected by waves because for long periods of the year (if not permanently) their coasts are protected by the formation of coastal sea ice.

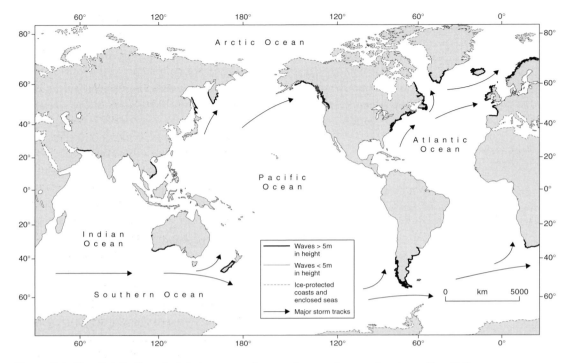

Figure 2.1 *The distribution of global wave environments: coasts experiencing > 5 m high waves are storm-affected.*
Source: from Briggs *et al.* (1997) (Figure 17.18, p. 319)

Waves

All waves possess alternating high crests and low troughs, and can be described (Figure 2.2) with reference to:

- **wave-height (*H*)**, the vertical distance between the wave trough and crest;
- **wave-length (λ)**, the horizontal distance between consecutive crests or troughs; and,
- **wave-period (*T*)**, the time interval between consecutive wave crests or troughs passing a fixed point.

Waves represent the transfer of energy, behaving differently depending on whether they are in deep or shallow water (Figure 2.3). In deep water, there is only minor movement of water in the wave. Although energy passes forward within a wave, water follows a circular pattern. A water particle may, from a starting position at the bottom of a wave trough, move up the face of an oncoming wave, against the direction in which the wave is travelling, to reach the crest and return to its original position. It will then travel down the back of the wave under gravity, this time moving forward with the wave, but finishing back at the bottom of the next wave

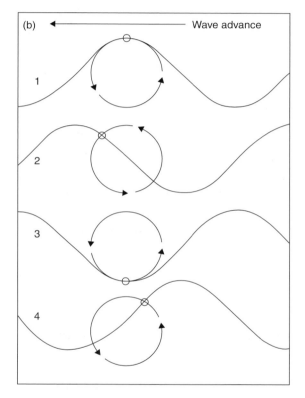

Figure 2.2 *(a) Wave description: wave-length (λ) is the horizontal distance (m) between consecutive wave crests or troughs, and wave-height (H) is the vertical distance between trough and crest; (b) the circular orbit of an oscillatory wave.*
Source: from Park (1997) (Figure 15.4, p. 407 and Figure 15.5, p. 408)

trough at its original starting position. This closed loop, where a water particle moves in a circular orbit, has given rise to the name **oscillatory wave**. The diameter of water particle orbits decreases with depth through the water column, so that eventually a depth is reached below which the water is unaffected by waves passing overhead. This depth is called the **wave-base**, and helps to define deep and shallow water waves, as deep water waves occur in water deeper than their wave-base, whilst shallow water waves are travelling through a water depth that is less than their wave-base. Wave-base can be established with reference to wave-length, and under various authors has been defined as a water depth equal to a range between one half and one quarter of the wave-length. Pethick (1984) employs the latter in defining wave-base, although he does state that some water movement can take place below this depth. The significance of wave-base cannot be overstated, as waves are perhaps the most important geomorphological agent on the coast. Waves can erode the coast, transport sediment, and ultimately deposit sediment, but all this activity takes place at and above wave-base. So in determining wave-base, we are going some way to determining the spatial extent of wave influence on a coast.

The study of waves involves some quite sophisticated mathematics, and although I will try avoid most equations, there are a few which are very useful to have to hand. The first of them introduced below helps us to establish the wave-length of deep water waves, for although the definition of wave-length is straightforward, its actual measurement at sea is an extremely difficult task. However, wave-length is related to wave-period, a wave characteristic that is far easier to determine. All that is required for an observer to establish wave-period is a fixed reference point and a stopwatch, recording the time taken for two

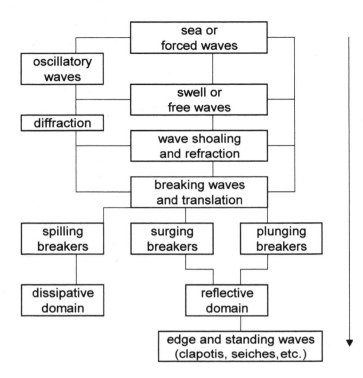

Figure 2.3 *Wave progression through deep to shallow water processes.*

wave crests to pass the fixed point. Wave-length can then be calculated using the following equation:

$$\lambda_o = \frac{gT^2}{2\pi}$$

where λ_o is the wave-length of a deep water wave, g is acceleration due to gravity (a constant at 9.81 m s^{-2}), T is wave-period in seconds, and π is the ratio of the circumference of a circle to its diameter, approximately 3.142. Of course, the significance of this equation lies in our ability now to calculate the all important wave-base from the wave-length. It also indicates that only modest increases in wave-period result in quite substantial increases in wave-length (Pethick 1984).

Another useful equation, which again incorporates wave-period values, determines wave-velocity or **celerity (C)**:

$$C_o = \frac{gT}{2\pi}$$

where C_o is wave-velocity (m s^{-1}) of a deep water wave. An observation that can be made from this equation is that waves with a high wave-period, and therefore, a long wave-length, travel faster than low wave-period, short wave-length waves (Suhayda and Pettigrew 1977; Pethick 1984).

Because we are dealing with coasts, all waves that are relevant to our study eventually encounter water depths shallower than wave-base (a depth equal to one quarter of the wave-length). When this happens the sea bed begins to interfere with the oscillatory motion of the deep water wave, and the wave starts **shoaling**, so becoming a shallow water wave (Figure 2.4). Wave-length and wave-velocity both decrease, but the energy from these reductions contributes to an increase in wave-height, which in turn leads to steepening of the wave-front. The circular orbits of the deep water wave become progressively elliptical, with the long axis of the ellipse parallel with the sea-floor, promoting alternating onshore-offshore currents

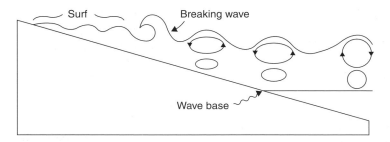

Figure 2.4 *Wave shoaling as an oscillatory wave enters shallow water with a transformation of water particles from a circular to elliptical orbit, leading to wave breaking.*
Source: modified from Briggs *et al.* (1997) (Figure 17.2b, p. 303)

that may be capable of picking up and transporting grains of sediment. The wave is now able to transmit matter as well as energy, and becomes a **translatory wave**. A shallow water wave will usually break when it encounters water depth that is less than wave-height (the **break-point**). In such shallow water, equivalent to less than one-twentieth of the wave-length, the equations for calculating deep water wave-length and wave-velocity can no longer be used, and are substituted by:

$$\lambda = T\sqrt{gd}$$

and

$$C = \sqrt{gd}$$

respectively, where d is depth (m).

Where an **incident wave** encounters wave-base along only part of the wave-front, the wave will undergo **refraction**. This situation may arise, for example, where submarine topography is variable, or where an incident wave approaches the shore obliquely with submarine topography descending uniformly offshore. Refraction is an important wave modification process because it affects the distribution of wave-energy along the shore (Figure 2.5). Shoaling promotes deceleration of wave-celerity, so that those parts of a wave in water shallower than the wave-base will slow, whilst sections of the wave still in water deeper than the wave-base will maintain its celerity. In this way the wave will bend or refract. On a map of refracting wave-fronts it is possible to draw lines called **orthogonals** normal to the wave-front and extending to the shore. Where orthogonals converge on a shore, wave-energy is concentrated, whilst where orthogonals diverge on a shore, wave-energy is dissipated. Therefore, along relatively straight shorelines with parallel and gradually descending submarine contours, there will be very little wave refraction and wave-energy will be uniformly distributed on the shore. However, along a highly indented coastline, characterised by headlands and embayments, wave-base is likely to be encountered off the headlands first. This would result in refraction of the wave around the headlands, covergence of the orthogonals and concentration of wave-energy on the headlands, and a corresponding divergence of orthogonals and dissipation of wave-energy in the bays. In this way, erosion may be expected to

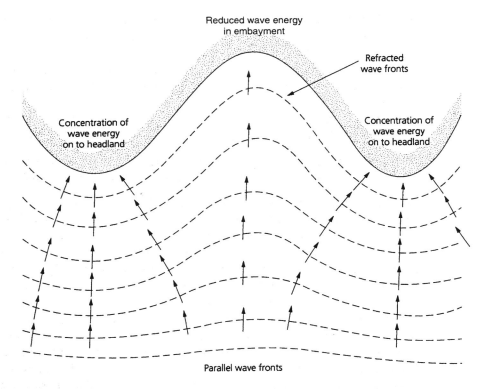

Figure 2.5 *Wave refraction on an irregular coastline: the arrows are orthogonals drawn at right angles to the wave fronts (dashed lines); energy is concentrated where orthogonals converge and dissipated where they diverge.*
Source: from French (1997) (Figure 2.3, p. 31)

dominate on the headlands, whilst depositional processes are more likely to characterise the lower energy embayments (McCullagh 1978).

Diffraction is another important wave modification process. When an incident wave encounters an obstacle in its path, such as an island or harbour wall, a wave shadow zone is created in the lee of the obstacle. However, as the wave passes the obstacle it is able to propagate or diffract into the shadow zone. It able to do this, because as a passing wave crest falls due to gravity, water is no longer confined by a continuous wave-front and so is allowed to escape or 'bleed' laterally into the shadow zone, creating waves that are smaller than the parent wave (Stowe 1996). Therefore, apparently protected shadow zone shores are still susceptible to wave activity and breaking.

The manner in which waves break against a shore is determined by the steepness of the incident wave, water depth and also the gradient of the shore (Galvin 1968; Summerfield 1991):

- **spilling breakers** tend to characterise shores with a low-angled gradient, regardless of wave-steepness, where water from the wave-crest spills or cascades down the wave-front producing lots of foam;
- **plunging breakers** occur on steeply inclined shores and are steep-fronted waves that tend to curl over and crash down on the shore, again producing lots of foam; and
- **surging breakers** are low waves that fail to break fully, the crest collapses, and the base of the wave-front advances up the shore as a rather foamless breaker.

Of course, these breakers occur along a continuous spectrum of breaker-types, and intermediate breakers have been described (e.g. collapsing breakers) (Figure 2.6).

The type of breaker is also associated with the way in which wave-energy is expended. Spilling breakers often roll in over relatively wide low-angled shores, upon which their wave-energy is dissipated gradually, giving rise to the term **dissipative domain**. Plunging and surging breakers, however, often breaking on steeper shores soon after encountering wave-base, will have a significant proportion of their energy reflected back out to sea as reflected waves, and hence operate in a **reflective domain**. Reflection of wave-energy may also occur off hard structures, whether natural (e.g. cliffs) or artificial (e.g. harbour walls). Reflected waves often travel at right-angles to the shore as **edge waves** and interact with incoming incident waves. Where the crests of both edge and incident waves intersect, the height of the waves combine to increase the wave-height at the intersection, and intersecting troughs combine to lower trough depth, resulting in both undulating wave-crests and troughs. Where reflected waves approximately equal the characteristics of, and travel in the opposite direction to, incoming waves then a **standing wave** known as a **clapotis** may be produced, in which the water goes up and down, but does not progress. Reflection occurring in a semi-enclosed body of

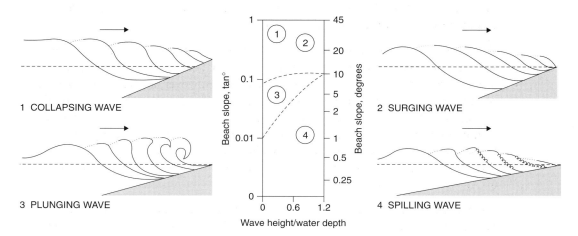

Figure 2.6 *Types of breaking waves as a function of wave height, water depth and beach slope gradient.*
Source: from Briggs *et al.* (1997) (Figure 17.3, p. 304)

Box 2.1

Coastal management implications of wave modification processes

As incident waves enter the nearshore zone they become modified by submarine topography, coastal configuration, and coastal substrate types, posing special problems for coastal managers and engineers, some of which are listed below:

- **Refraction.** Rises and mounds on the sea-floor may focus wave-energy on to certain stretches of the coast. For example, artificial hard-standing for piers may focus wave-energy on to the coast where the pier is connected to the shore, which in extreme cases may undermine foundations and result in pier detachment from the shore. Also, the famous case of the Long Beach breakwater in California is worth recounting, for in April 1930 it was badly damaged by long wave-length waves from a southern hemisphere storm that were refracted by a 'bump' in the continental shelf as the wave-train passed over it. The wave-energy was so well focused that the coasts of neighbouring Long Beach were completely unaffected (Stowe 1996).
- **Diffraction.** This modification process allows waves to enter into areas perceived to be protected, such as the lee-side of islands, behind breakwaters and within harbours. Island shadow zones are often characterised by choppy wave conditions as waves diffract in from both sides of the island, sometimes creating conditions more hazardous to sailors than exposed coasts, and makes island hopping quite uncomfortable for those without sea-legs! Also, harbours and breakwaters should be designed to minimise diffraction by aligning structures parallel to the fetch direction, as far as is possible, to protect valuable water-craft and sea-side buildings from storm wave attack.
- **Reflection.** This modification process transforms incident waves into reflected waves. The formation of reflected waves in confined coves and harbours can create standing waves which may lead to collision and damage of closely-moored boats, and produce unsuitable sailing conditions. Therefore, harbour walls should be designed to dissipate and absorb wave-energy rather than reflect it. This is largely achieved by building porous structures, such as rubble-mound barricades and boulder amouring. There is a further advantage to minimising reflection in that reflection can create resonance within the reflecting structure, which may lead to the development of internal weaknesses, degradation and the ultimate replacement of the structure at cost.

water, such as a cove or harbour, may produce a **seiche**, another type of standing wave, which sloshes back and forth across and out of the water body (Stowe 1996).

Wind waves

Wind waves are those created by the friction that arises when wind blows over the sea surface (Figure 2.7). Both air and water are fluids, but with different densities, and as such frictional drag forms waves at the interface between them. Therefore, wind waves are indirectly created by solar energy and categorised as exogenetic.

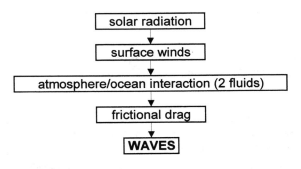

Figure 2.7 *The creation of wind waves.*

Many waves are generated by localised storm activity at sea, and whilst the waves are still within the generation area they are termed **sea waves** or **forced waves**. Upon leaving the generation area these waves lose height and energy to become **swell waves** or **free waves**, which may travel many thousands of miles before breaking on a distant shore. This swell wave reduction in height and decrease in energy occurs rapidly to begin with, so that an initially 10 m high wave may in its first 200 km of travel from the generation area be reduced to 2 m in height with a corresponding 80–90 per cent energy decrease, but thereafter will diminish only slightly (Komar 1976; Summerfield 1991).

The height of a wind-generated wave is closely associated with wind velocity. Observations have shown that wave height is proportional to the square of wind velocity, so that higher wind velocities create higher waves and generally rougher seas. This relationship can be summarised as:

$$H = 0.031 \times U^2$$

where H is wave-height (m), U is wind velocity (m s^{-1}), and 0.031 is an empirically derived constant. Wave height is also affected by **fetch**, which is the distance over which wind interacts with the sea-surface to create and propagate waves (Plate 2.1). Observations indicate that wave height is proportional to the square root of the fetch, so that higher waves are associated with a long fetch. This relationship can be summarised as:

$$H = 0.36 \times \sqrt{F}$$

where H is wave-height (m), F is fetch (km), and 0.36 is an empirically derived constant. The combined affect of high wind velocities and a long fetch result in high waves, so that extensive seas like the Atlantic and Pacific Oceans, for example, experience waves in excess of 25 m high (Pethick 1984; NERC 1991). When discussing wave-height, however, it is often more meaningful to refer to **significant wave-height (H_s)**, which is the mean of the highest one-third of waves affecting a coast. For example, significant wave-height in the Atlantic Ocean is approximately 2 m.

In a wave-generating area, wind does not create uniform groups of waves, especially during storms, but more often creates a chaotic assemblage of waves with varying wave-lengths and periods. As we discussed above, waves with a long wave-length travel faster than waves with a short wave-length. From this it is clear

Plate 2.1 *The influence of fetch on waves: (a) long fetch waves on the eastern Atlantic coast at Bretignolle-Sur-Mer (western France) characterised by long wave-lengths and high wave-heights; and (b) short fetch waves on the northeastern coast of Australia at Yorkey's Knob (near Cairns, Queensland), where the Great Barrier Reef prevents swell waves from the Pacific Ocean entering the coastal waters, thus these waves are created landward of the reef and are characterised by short wave-lengths and low wave-heights.*

that swell waves leaving a generating area will eventually sort themselves with the long wave-length waves in the front, and the short wave-length waves at the back, of the travelling wave train. This process is called **dispersion** and is responsible for many Atlantic and Pacific coasts (e.g. southwest Britain, northwest France, California, and New South Wales) receiving long wave-length waves suitable for surfing. Whereas within enclosed seas, such as the Mediterranean, dispersion is limited by the restricted fetch, resulting in the co-occurrence of long and short wave-length waves that produce choppy wave conditions (Pethick 1984; Stowe 1996). Where two groups of waves travelling in the same direction with similar wave-lengths coincide (**superposition**), but perhaps slightly out of phase with each other, they may produce a wave phenomenon known as **surf beat**. As with edge waves discussed earlier, where crests of waves in both wave groups coincide, the heights of the individual waves combine to increase wave height. However, the combined effect of a wave-trough in one group coinciding with a wave-crest in another group is to reduce wave-height. Therefore, higher than average waves are formed where the two wave groups are in phase, producing ideal surfing conditions, but lower than average heights are achieved where the wave groups are out of phase.

Tsunami waves

Tsunami are waves that are produced by a number of different mechanisms, including displacement of the sea-floor by movement along a fault (associated with an earthquake) or a submarine slide, volcanic activity and asteroid impact. They have also been called tidal waves and seismic sea waves, but these are inappropriate as tsunami are unrelated to tides and those created by asteroid impacts are cosmogenic rather than seismogenic. The word tsunami itself is not ideal, as it is Japanese for harbour wave, and these waves are not confined to harbours. In the case of sea-floor displacement, sea level falls directly above the area of subsidence as water infills the newly created space (Figure 2.8). Surface water then rushes into the area to restore sea level, but overcompensates creating a localised bulge in sea level which subsequently propagates outward as a tsunami (Smith and Dawson 1990). Asteroid impacts act in the same way as throwing a stone into a pool, where a combination of displacement and energy transfer occurs between the impacting body and the surface water.

In the open ocean, tsunami are similar to wind waves in that wave heights are often small, commonly no more than 1 m high, and ships at sea would not notice a tsunami passing beneath them. However, tsunami possess extremely long wave-lengths, measured in hundreds of kilometres, which means that they may almost always be considered a translatory wave and that they travel at extremely high speeds, for example 600 km h^{-1} in a water depth of 3,000 m (Summerfield 1991). In this way, tsunami are able rapidly to cross large oceans, such as the Pacific, an ocean that is particularly prone to tsunami with at least one tsunami event occurring every twenty-five to fifty years in Hawaii. As with all waves, when tsunami enter progressively shallow water, usually as they encounter the continental shelf, their velocity decreases and wave-height increases, often with catastrophic consequences for coastal populations. They may also refract, diffract and reflect, further concentrating their energy on particular parts of the coast, affect areas in the shadow zones of islands and create edge and standing waves respectively.

Although a tsunamigenic event will usually create not one tsunami, but a series of tsunami waves that may repetitively affect a coastline for up to twenty-four hours, these events occur only occasionally, even on tsunami-prone coasts. Summerfield (1991) considers tsunami to be of such low frequency that they are of less significance than high frequency wind waves in contributing to the physical development of coasts. However, Bryant *et al.* (1996) suggest that along some coastlines, such as New South Wales in Australia, tsunami do make a substantial contribution to coastal evolution, and that coasts should be evaluated individually as to the extent tsunami may have on their evolution. After all, the geomorphological significance of low frequency, but high magnitude events in other physical systems, such as river flood events, has long been appreciated.

Erosional coasts

The input of energy, whether derived from solar, seismic or cosmic sources, into a coastal system via waves constitutes one of the main forces driving

Figure 2.8 *The formation of a tsunami wave as a result of seafloor displacement i.e. submarine slide: stage 1 – submarine slide occurs; stage 2 – a depression forms in the sea surface above the slide; stage 3 – surrounding water rushes in to fill the depression and piles up (an exceptionally 'low tide' may be seen on shore during this stage); stage 4 – the piled up water now flows outward forming tsunami waves.*
Source: reproduced with permission from Smith and Dawson (1990)

Box 2.2

Tsunami of 17 July 1998, Papua New Guinea

As night was falling on the evening of 17 July 1998 along the coast of West Sepik Province in northwest Papua New Guinea, the inhabitants of fishing villages felt their homes tremble – that was at 6.49 p.m. local time. At 6.50 p.m. the villagers heard a roar, like a jet-plane, and saw the sea rise and come towards them, not as a stereotypic towering wall of water as portrayed in Hollywood movies, but as a swell that rose to engulf them. Within the space of a few minutes three tsunami had struck the coastline, the largest estimated to be some 10 m (30 ft) high. A fourth but less violent wave arrived later, and by 7.07 p.m. the sea was calm once more. However, the coast had been battered severely by tsunami, leaving a trail of devastation along a 30 km stretch of coast west of Aitape. Two of the worst hit villages were situated on a narrow spit that separated Sissano lagoon from the sea; Warapu (population *c.* 1,800) and Arop (population *c.* 2,000) were completely destroyed, no buildings left standing, palm trees ripped out, and men, women and children washed out to sea. By 15 August 1998 the death toll was estimated at over 2,200.

The cause of the trembling felt by the villagers prior to the arrival of the tsunami was an earthquake with an epicentre some 19 km (12 miles) offshore at a crustal depth less than 33 km, measuring 7.1 on the Richter scale. This probably stimulated vertical displacement along a fault and/or triggered submarine slides (González 1999). The speed at which the first tsunami reached the coast was due to the close proximity of the coast to the epicentre, in fact the coast was within the general tsunami-generating area. The severe height of the tsunami may have been partly due to the occurrence of deep water close to the shore, with a steep gradient leading to the coast helping to pile up the water. Ironically, this whole event was precursed on 4 July 1998 by another earthquake in the region, measuring 5.5 on the Richter scale. The reasons for inaction regarding this precursor event may lie in the perception that this area was considered of low tsunami risk, with the last major event occurring in 1907, and so largely forgotten.

The close arrival of tsunami soon after the earthquake gave no time for evacuation in this case, but on similar spit coastlines with little elevated ground, in situations where tsunami may only be a few metres high and have longer travel times, thus providing some warning, the best precaution perhaps is to build elevated refuge platforms. It has been argued that many deaths would have been prevented in 1979 on San Juan Island, a Colombian spit with a maximum altitude of 3 m above sea level, if residents could have got above the 3 m high tsunami that struck.

physical coastal processes. When considering coastal processes and resultant landforms, it is logical to begin with erosional processes and landforms, as it is erosion that is largely responsible for providing sediment for depositional processes, and in this way depositional coastal systems are to a significant extent dependent on mass input from erosional systems.

The ability and extent to which waves may physically erode a coastline are dependent on three sets of variables:

- Wave environment of the coastline. This includes orthogonal fetch direction, significant wave-height, frequency of high magnitude wave events, such as storms and tsunami.
- Geology of the coastline. This includes rock-type or **lithology** with its inherent hardness and susceptibility to physical and chemical weathering, geological structure (bedding planes, joints, faults and folds), provision of rock suitable for use as erosional tools by waves, and inherited geological weakness, such as weathering associated with past climates and former sea level situations.
- Morphology of the coastline. This includes areal configuration and inherited topography (headlands and bays promoting refraction), cliff heights and slope angles, and submarine topography (again influencing refraction).

In nature, however, it is not only waves that erode coastlines, but also tidal action, subaerial conditions (e.g. climate), and biological activity, including the activity of humans, may make a contribution, which will be discussed later.

The mechanisms of wave erosion are varied and have attracted the attention of many authors who have established a number of terms to express these processes. Waves breaking on to bedded, jointed or faulted rocks can create **hydraulic pressure** in these structural voids, which may lead to weakening and readying of rocks for detachment by the process of **quarrying**, where a wave removes loose blocks. Detached blocks may break down under wave activity through processes of **attrition**, such as **abrasion**, which involve the rubbing of rocks against one another, gradually rounding the rocks and reducing their size. Sediment provided by quarrying and attrition may then be used as erosional tools by waves to further erode rock through **corrasion**, which is the mechanical weathering of rock surfaces by abrasion.

Cliffs

Perhaps the most well known of all coastal erosional landforms are cliffs, which may loosely be defined as any steep coastal slope affected by marine processes, although subaerial processes also play a significant role in cliff formation. The form that a cliff takes may be determined by:

- Inherited characteristics. The sea may rework steep slopes initially formed by non-marine processes under different sea level situations, for example, some **plunging cliffs** that rise abruptly from deep water were originally the sides of now submerged glaciated valleys.
- Wave activity. Sufficiently energetic waves are required not only to erode rock, but also to remove debris created by wave erosion and material deposited at the foot of the cliff by subaerial processes operating higher up, which would, if not removed, obscure and protect the cliff, restricting its development.

● Geology. The hardness and structure of the rock are very important in cliff development, as some soft rock may fail to support steep slopes altogether, regardless of the level of wave energy. Also, rocks with bedding planes that incline or dip seaward may slip off into the sea, with the resultant cliff angle determined largely by the dip angle, whilst rocks that possess horizontal bedding or dip landward may be capable of developing near-vertical cliff faces (Figure 2.9).

Cliff retreat often proceeds through the formation of a **wave-cut notch** at the base of the cliff, which effectively undermines the cliff leading to slope failure, either in the form of rotational slumping or vertical cliff collapse (Plate 2.2). The notch itself is formed through quarrying and corrasion, and is best formed where waves actually break on to the cliff. Notches are unlikely to form on plunging cliffs, which simply reflect swell waves that it intercepts, or on cliffs fronted by a wide and low-angled shore which is capable of dissipating most wave-energy before it actually reaches the base of the cliff. Therefore, cliffs fronted by a narrow shore are likely to possess notches, and retreat in a cycle of notch formation, cliff failure and debris removal.

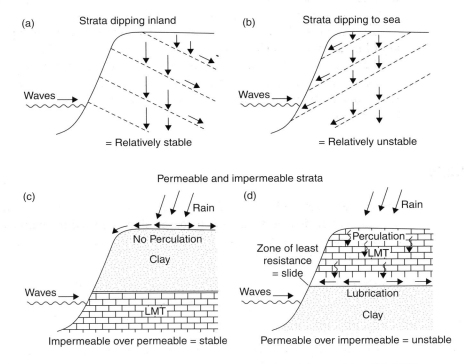

Figure 2.9 *Influences of geological structure (a and b) and lithology (c and d; LMT = limestone) on coastal cliff development.*
Source: from French (1997) (Figure 2.8, p. 40)

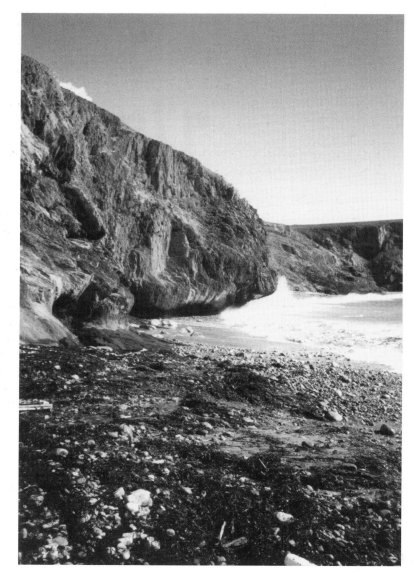

Plate 2.2 *A wave-cut notch actively forming at the base of cliffs at Pointe de Dinan on the Crozon Peninsula (Brittany, France).*
Photograph: kindly supplied by Dr Janice Ross (Bath Spa University College)

Other mechanisms of cliff retreat include the activity of landslides, mudflows and rotational slumps (Figure 2.10a) in soft and weak lithologies, moving material from high up the cliff to the base in an attempt for the cliff slope to attain equilibrium. However, the subsequent removal of this material from the base of the cliff by wave activity destabilises the cliff and stimulates further mass movement, sustaining retreat. Also, cliff retreat can perpetuate itself through the development

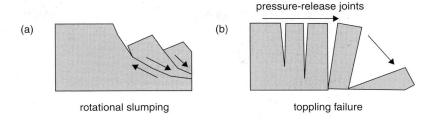

Figure 2.10 *Mechanisms of cliff failure: (a) rotational slumping (arrows indicate relative movement); (b) toppling failure.*

of **pressure-release jointing**. This occurs because rock expands as the confining pressure created by surrounding rock decreases as cliff retreat proceeds. Expansion takes place in the same direction as pressure release and so joints open up parallel to retreating cliff faces. Debris, water, and organisms can invade these joints helping to prise them open, eventually leading to **toppling failure** (Figure 2.10b), when large slabs of rock topple onto the shore.

Variations in rock strength along the length of a cliff can often lead to **differential erosion**, where weaker rocks are eroded at a faster rate compared to relatively stronger rocks. This may result in the formation of caves, arches and stacks. Caves are progressively excavated through the combined action of hydraulic pressure, quarrying and abrasion exploiting a weakness in the rock, such as a fault. A bridge-like arch is formed where a cave develops to such an extent that it emerges out the other side of a headland. Whether during the cave or arch stage, the ceiling may become unstable through wave action and/or gravity, and collapse. If ceiling collapse occurs gradually, with vertical holes appearing as ceiling sections fall at different times, then **blowholes** may be formed, which may issue a cliff-top water geyser under favourable wave conditions. A stack is created upon the complete destruction of an arch's ceiling, so that it is no longer attached to the main cliffline and is essentially an island (Figure 2.11).

Shore platforms

A consequence of cliff retreat is the creation of intertidal **shore platforms** which extend seaward from the base of the cliff. Although shore platforms are initially created by wave quarrying and abrasion activities, it is now appreciated that continued platform development is often aided by **bio-erosion** and weathering, particularly where tidal exposure is significant, and hence the abandonment of 'wave-cut platform' as a descriptor for these landforms. Also, shore platforms are seldom horizontal, and often possess a gentle seaward slope of up to 3°, sometimes with small cliffs around the low-tide level and relatively steep ramps at the high-tide level. Rock or tide pools are also common on shore platforms, being contained within hollows excavated by quarrying and weathering.

Figure 2.11 *The geomorphology of a rocky coastline.*
Source: from Briggs *et al.* (1997) (Figure 17.14, p. 316)

Box 2.3

Beachy Head – cliff collapse in southern England

Beachy Head is a well-known chalk cliff along the Sussex coast near Eastbourne in southern England. On the night of Sunday 10 January 1999, a very large section of the chalk cliff collapsed. The actual dimensions of the cliff section that collapsed is uncertain; the Environment Agency (1999) suggest a 150 m high slab comprising hundreds of thousands of tonnes of chalk, but an item in the Geological Society of London's magazine (Nield 1999) reports a more conservative estimate of a 60 m high and 15 m thick cliff section. Whatever the dimensions, it was certainly a significant size and the fall created a causeway almost linking the coast with Beachy Head's offshore lighthouse.

The Environment Agency (1999) speculate that a series of climatic events may have contributed to the collapse:

● for 30 months prior to the autumn of 1998, southern England suffered a drought which may have weakened the chalk through drying;
● then through the autumn and winter of 1998 continued heavy rainfall is thought to have saturated the chalk;
● severe storms affected the coast of southern England for twelve days over the Christmas period, perhaps weakening and undermining the base of the cliff; and
● cold conditions followed the storms with sub-zero temperatures and freezing within the chalk may have produced further internal stress that provided the trigger for collapse.

It is highly probable that these climatic events contributed in some way to this cliff collapse, but these cliffs have been actively retreating since the rise of post-glacial sea level, so that the scale of this particular collapse is unlikely to be unique in the post-glacial erosion history of Beachy Head, and in my opinion does not reasonably constitute 'solid evidence that climate change, which was predicted as a result of global warming, had arrived with a vengeance', as reported by the Environment Agency (1999).

The dimensions of shore platforms, under stable sea level conditions, are intimately related to wave processes (Figure 2.12a). Platform width, although associated with the rate of cliff retreat, has a finite limit (Trenhaile 1999). This is because as a quasi-horizontal platform develops in front of a cliff, it increasingly acts to dissipate wave-energy before it reaches the base of the cliff, and thus beyond a critical platform width waves are unable to erode or remove debris protecting the cliff face, halting platform expansion. Under such circumstances, the only way in which cliff retreat can be maintained is if the height of the platform is reduced relative to sea level, so diminishing its dissipative effect. However, there is also a limit to platform lowering, because shear stress between the platform and passing waves decreases with water depth, so that at a critical depth wave activity becomes unable to erode the platform. Tidal range also has a part to play, with the potential to develop wider platforms under increasing tidal ranges (see pp. 78–83).

A number of shore platform classification schemes have been proposed. They are often classified according to their position within the tidal frame. This has given rise to quasi-horizontal low-tide and high-tide platforms, or intertidal forms which slope from a high-tide level to low-tide, perhaps terminated by a low-tide cliff. Alternatively, shore platforms have been categorised by morphology, whether sloping, quasi-horizontal, or near-vertical (Types A-C of Sunamura 1992) regardless of tidal position (Figure 2.12b; Plate 2.3).

Ecology of rocky shores

Rocky shores represent quite an extreme environment, characterised by:

- Continual erosion of cliffs and platforms. Provides dynamic and only very temporary habitats – for example, nesting seabirds and salt-tolerant flowering plants colonising cliffs regularly have to re-establish themselves after cliff-falls.
- Wave activity. This makes attachment a problem which has led to the development of clinging strategies for many species of algae and invertebrates, although even these might fail in severe wave conditions. Wave activity also largely prevents the accumulation of nutrients, and most organisms here extract their nutrient requirements directly from sea-water.
- Highly variable environmental conditions. This particularly concerns salinity, temperature and pH ranges, and water, oxygen and light availability. Tidal

(a)

(b)

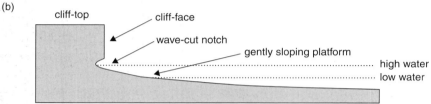

Type A (sloping) shore platform profile

Type B (quasi-horizontal) shore platform profile

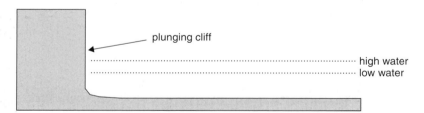

Type C or plunging cliff profile

Figure 2.12 *(a) Negative feedback system between wave energy, cliff retreat and platform expansion: high wave energy leads to high rates of cliff retreat, but consequent platform expansion increases dissipation, reducing wave energy at the cliff and restricting further retreat; (b) profiles of shore platform morphology types.*

Plate 2.3 *Types of shore platforms: (a) Type A or sloping shore platform and cliffs in well-bedded alternating lias limestone and shales (Lower Jurassic) of the Glamorgan Heritage Coast (Wales, UK); and (b) Type B or quasi-horizontal shore platform at Coledale Beach south of Sydney (New South Wales, Australia).*

factors, especially drying/wetting (dessication), freshwater seepage from cliffs reducing salinity, and water **turbidity** and depth at high tide limiting light are important.

The distribution of organisms on shore platforms is often zoned (Table 2.1). Because many intertidal species inhabiting these shores gain nutrients from the water, productivity is greatest near low tide. Here species richness, community complexity, and competition are high, but all decrease towards the high-tide level. Platform morphology influences this tide-related zonation, as a high platform slope angle would accentuate a zonation, whilst environmental conditions on a horizontal platform may remain spatially uniform at all states of the tide, negating zonal establishment.

Coral reef coasts

It is logical to consider coral coastlines here because the majority of reefs are composed of dead coral limestone, as only a thin layer of the reef surface supports living coral. This coral limestone is created by the activity of both small animals called polyps, that build delicate limestone structures that become the coral colony, and microscopic coralline algae that cement the delicate coral structures into a hard limestone pavement. Dead coral is susceptible to erosion by wave activity, which can reduce the limestone to a rubble that may be transported by wave-induced currents to infill active reef structures, or to create rubble mounds suitable for new coral colonisation, or to be swept up into a pile to create coral islands or **cays**. Therefore, wave activity plays an important role in coral reef erosion, deposition, and general morphological development.

Many of the world's coral reef systems have existed for millions of years; for example, parts of the Great Barrier Reef on the eastern seaboard of Australia have existed for 18 million years. However, their development has been influenced by the rise and fall of sea levels through **Quaternary** interglacial and glacial stages respectively (Larcombe and Carter, 1998; also, see p. 138). During sea level highstands coral building on the continental shelf has occurred, but during lowstands the continental shelf becomes terrestrialised and the reefs become limestone hills. Limestone is prone to dissolution and **karst** features, such as caves, develop at these times. Under subsequent high sea levels, corals re-establish themselves, leading to cyclical reef evolution through time.

There are many forms of coral reef, however, there are three main types:

● **Ribbon reefs.** These are generally very narrow coral reefs that often occur on the seaward edge of reef areas. In the northern Great Barrier Reef, ribbon reefs develop on the edge of the continental shelf and may represent the upward growth fringing reefs from a previous sea level lowstand.

Table 2.1 The zonal distribution of characteristic organisms (lichens, algae, molluscs, and some other organisms) in relation to tide levels on a rocky shore (Cremona, 1988)

Tide levels	Zone	Environment	Lichens	Algae	Molluscs	Other organisms
	Splash zone	High salinity, temperature extremes, dessication, limited water	Orange (Xanthoria parietina and Calo placa marina), grey (Ochrolechia parella and Lecanora atra), and green lichens (Ramalina siliquosa)	Prasiola stipitata	Littorina neritoides	
High tide	Upper shore	Wide temperature variation, drying out at low tide	Black lichen (Verrucaria maura)	Pelvetia canaliculata, Fucus spiralis, Lichina pygmaea, Enteromorpha intestinalis, and Porphyra umbilicalis	Littorina saxatilis to L. rudis	Acorn barnacles
Mean tide	Middle shore	Variable temperature, light limited at high tide		Ascophyllum nodosum, Fucus vesiculosus, Cladophora rupestris, Ulva lactuca and Ceramium rubrum	Gibbula umbilicalis, Littorina littorea, L. Littoralis, L. obtusata, Patella vulgata, Mytilus edulis and Nucella lapillus	Actinia equina
	Lower shore	Stable environment, high competition, light limited by water depth and turbidity at high tide, high wave activity		Fucus serratus, Corallina officinalis, Lithophyllum spp., Himanthalia lorea and Codium tomentosum	Patella vulgata and Nucella lapillus	Halichondria panicea, Anemonia viridis, and Bunodactis verrucosa
Low tide	Sublittoral	Stable environment, high wave activity, light limited		Laminaria spp., Chondrus crispus, Gigartina stellata, Rhodymenia pseudo-palmata, and Saccorhiza bulbosa	Gibbula cineraria, Calliostoma zizyphinum and Patina pellucida	

- **Fringing reefs.** These reefs develop along the coast of the mainland and of continental islands, and are amongst the only reefs accessible from the land without the use of a boat, and consequently are vulnerable to visitor pressure.
- **Patch reefs.** These are often oval to round reefs that occur on the continental shelf and develop through a series of six stages: (1) an ancient reef platform is re-activated by rising sea level to produce (2) a submerged reef that develops into (3) irregular reef patches around the platform periphery; these patches merge to create (4) crescentic reefs that eventually form a circle (5) enclosing a lagoon; sediment infilling of the lagoon leads to the formation of (6) a planar reef surface upon which a cay island may develop. The vast majority of coral reefs on the Great Barrier Reef are of this type.

The morphology of these reefs usually comprises an eroded fetch facing 'windward' side and a relatively protected 'leeward' side characterised by the development of new coral colonies, often with massive stacks called 'bommies' which are composed of *Porites* coral (this does not apply to fringing reefs that have landward sides that adjoin the coast). Between the windward and leeward/landward sides, a reef platform develops that may often possess lagoons. Debris derived from erosion of the reef on the windward side is often deposited on the platform surface, graded from coarse to fine debris towards the leeward side. Wave refraction on subrounded patch reefs can transport this debris to the point on the reef platform where the wave orthogonals intersect, sometimes accumulating enough debris at these locations to begin the process of cay formation (Figure 2.13).

Coral cays

Coral cays develop as a consequence of sediment build-up on a reef platform, perhaps initially as nothing more than a sand or shingle bank. However, as sediment continues to accumulate through wave and also tidal action, it may reach a point at which it is submerged during only one or two high tides a year. At this point, seabirds may begin to nest on the cay bringing seeds with them, to

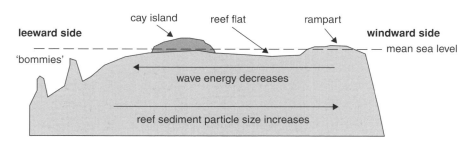

Figure 2.13 *The geomorphology of a coral patch reef.*

complement those arriving by floating from mainland or continental island sources, and guano helps fertilise the cay sediment. Dune and mangrove vegetation may then develop, contributing organic matter to the developing cay soils and also trapping and allowing sediment to accumulate above high tide, leading to further cay expansion. When the cay attains a suitable size, the sandy cay substrate may accommodate a freshwater reservoir that enables halophobic (salt-intolerant) plants, such as trees, to become established.

Vegetation helps to stabilise the cay, but this is also aided by the development of relatively hard and cemented cay sand. This may develop in two ways, either:

1 by the downward percolation of precipitation of phosphate from seabird guano to produce a hard pan called cay sandstone, which is sometimes of economic importance; or
2 the precipitation of calcium carbonate in the pore spaces between the sand grains, so cementing them together to form **beachrock**.

These cemented sands underlie cays and serve to increase cay stability. However, at all stages of cay development, the cay is vulnerable to severe wave activity, such a storm waves associated with tropical cyclones. In many cases, cays have been destroyed by a single storm event, but quite often the cay is able to respond to short- and long-term wave variations by changing its morphology, eroding in one place and depositing in another. Therefore cays, like many coastal landforms, are considered to be **morphodynamic** structures, being able to change morphology in response to the dynamics of their environment. In this way, cays are often shifting, seeking to obtain equilibrium with the prevailing wave environment at a given time. In the field, morphodynamic changes may be detected by the presence of a number of criteria (Plate 2.4), including:

● the exposure of beachrock around cay margins indicates the loss and removal of overlying unconsolidated cay sand by wave erosion and transport;
● the presence of cliffs (usually ≤ 1 m high) backing cay beaches indicates that wave erosion and removal of unconsolidated cay sand by wave transport are actively occurring;
● the occurrence of fallen trees with exposed roots on beaches indicates active undermining of cay vegetation by wave activity (this is usually seen in conjunction with cliff formation).

In contrast to reef build-ups, which may be several million years old, coral cays are only a few thousand years old. This is because unconsolidated cay sand is not easily preserved between sea level highstands. Also, within any given interglacial highstand, cays develop best after the stabilisation of sea level and the expansion of reef platforms. This is because under rising sea levels reef growth is primarily vertical, attempting to keep pace with sea level; however, horizontal growth is stimulated when sea level stabilises. Subsequently, reef platform expansion provides the material and foundation for cay development.

Plate 2.4 *Exposed beachrock and fallen trees, indicators of coral cay instability at Green Island, Great Barrier Reef, Australia.*

Coral reef ecology

Although polyps are animals, the availability of light is of crucial importance in the formation of reefs. This is due to the symbiotic relationship between polyps and tiny green algae, called zooxanthellae, that they play host to. Polyps feed by catching microscopic floating prey with their stinging tentacles; however, nutrients are supplemented by photosynthesising zooxanthellae, that also utilise some polyp waste products, such as carbon dioxide. Corals occur at all latitudes of the world, but it is only in relatively warm waters of 18°C and higher (ideally around 26°C) that they are able to construct reefs. It is the combination of warm water, normal salinity, zooxanthellae and good light availability for their photosynthesis, that promotes high rates of coral production and reef development.

The symbiotic relationship between animals and plants on reefs is not restricted to coral polyps and zooxanthellae, as clams, sponges and sea squirts also have symbionts. Also, nonsymbiont algae proliferates on reefs and is the food source for planktonic and benthonic invertebrates, and also for higher order animals such as fish and turtles. These herbivores may then be preyed upon by carnivores. Photosynthesis in both symbiont and non-symbiont plants, therefore, forms the energy basis of the reef ecosystem. Bacteria consume organic wastes and dead plants and animals, which in turn are consumed by filter feeders, such as sponges and clams.

Some animals actually prey on the coral, sometimes with serious consequences. The crown-of-thorns starfish (*Acanthaster planci*) is one such predator, growing up to 80 cm in diameter, with up to twenty-one arms covered with venomous spines. It is normally an uncommon animal inhabiting Indian and Pacific Ocean reefs; however, outbreaks occasionally occur when their populations explode to plague proportions, often numbering thousands to millions of individuals per reef. Under such circumstances an entire reef might be completely destroyed in two to three years. The reasons for such outbreaks are not fully understood and may be due to either natural factors, human activity or a combination of the two. The rate of reef recovery following an outbreak is variable: staghorn (*Acropora*) corals with a growth rate of 20–30 cm a year may recover within twenty years, but massive corals with a much slower maximum growth rate of 4 mm a year will take considerably longer to recover.

Coral reef ecosystems are extremely sensitive to environmental stress. A common sign of coral stress is 'bleaching' (Brown and Ogden 1993). This apparent bleaching is brought about by the coral polyps ejecting zooxanthellae, and may be in response to:

- Water temperature variations below and above coral tolerance levels. This may be a particular problem during pan-Pacific El Niño events that increase surface water temperatures in some regions.
- Salinity variations brought about by increased freshwater run-off via rivers and streams into coastal areas of coral reefs. Seaward reefs may also be affected where substantial rivers in flood jet freshwater out on to the continental shelf.
- Increased turbidity due to high concentrations of suspended silt limits light availability for zooxanthellae photosynthesis. This problem is often linked to increased river run-off in association with changing land practices, such as deforestation and agriculture, that promote soil erosion. Conversely, ultra-violet stress may also cause bleaching in non-turbid waters.

The increased frequency of bleaching events in recent years has led to speculation that global climate change may be partly responsible, and indeed the possible influence of climatic factors, such as El Niño and changes in some regional precipitation regimes, offers some support to this argument.

Barrier islands

Barrier islands are depositional offshore linear features, separated from the mainland by a lagoon and orientated parallel to the coast. They are usually comprised of sand that is built above the high-tide level and stabilised by vegetation. They are very variable in size from tens to hundreds of metres wide, hundreds to thousands of metres long, and may support sand accumulations up to 100 m high. Typically, barrier islands occur on coasts with a low gradient and low

Box 2.4

Impact of tourism on coral cay reefs – Green Island, Great Barrier Reef

Coral cays are often sites of tourist activity, which creates pressures that may affect cay morphology and influence its stability and vulnerability. Australia's Great Barrier Reef, a World Heritage Area, attracted 1.5 million tourists in 1994–1995, who spent $1 billion, representing a ten-fold increase of numbers since the early 1980s. This increase in tourism, much of which is centred on cays, is largely due to the introduction of fast catamarans that are capable of rapidly ferrying tourists from the mainland, making day trips to the reef possible. Green Island, lying on a patch reef near Cairns, is one of the most popular cay resorts. A survey of Green Island since 1936 has shown a number of morphological and environmental changes:

- The position of a spit located at the western end of the cay has shifted from the northwest in the 1940s, to the southwest in the 1950s, and back again to the northwest in the 1970s where it has remained. It is thought that this has little to do with tourist pressure, but more concerned with cyclone activity, particularly cay streamlining in response to the prevailing cyclone direction.
- Also on the western end of the cay, a groyne was constructed to protect the resort beach. However, this interrupted longshore drift patterns and reduced new sand input to the western beach, leading to localised erosion. In response to the erosion, the groyne was later dismantled, but then rebuilt following the artificial emplacement of 18,000 m^3 of fine sand on to the beach by the Beach Protection Authority between 1974 and 1976. It appears that by 1978, however, this sand had been lost from the beach to the relocated spit on the northwest corner of the cay.
- The area of the reef flat covered by algal mats and sea-grasses (e.g. *Halodule, Halophilia, Symodocea* and *Thalassia*) increased from 900 m^2 in 1945 to 130,000 m^2 in 1978, and has been interpreted as an effect of **eutrophication** from sewage released directly on to the reef flat. Indeed, algal and sea-grass colonisation was first apparent around sewage outlets. This is a problem not only for Green Island, but also many other cay resorts and fringing reefs close to mainland sewage outfalls. Measures to limit effluent pollution on Green Island have been in place since 1994 with the construction of a tertiary sewage treatment plant. Approximately 30 per cent of the treated effluent, that was previously discharged into the sea, is now used to irrigate the grounds of the resort, but careful monitoring ensures that the island's natural aquifer is not contaminated. Also, water conservation is being promoted so that the amount of waste water produced is reduced. For example, the installation of water-efficient toilets and showers in the tourist resort has reduced water consumption from 74 litres per person per day in 1995–1996 to 55 litres in 1996–1997, which in total is a change from approxmately 21 to 15 million litres.
- The total volume of sediment contained within the cay appears to have decreased and is thought to represent a negative feedback to generally increased eutrophy within the cay system. This is due to the sediment-trapping ability of sea-grasses which interrupts the normal sediment exchange between the reef flat and the cay. Sediment is often transferred from cay to reef flat during cyclones, but returned during fairweather wave conditions. Increased sea-grass colonisation on the reef flat traps and prevents this sediment being returned to the cay. Indeed, in 1998 approximately two-

> thirds of the Green Island periphery was characterised by exposed beachrock, back-beach cliffs and fallen uprooted trees indicating cay instability.

tidal range, and thus over 10 per cent of the world's coasts have developed barrier islands, including the Atlantic coast of the USA and on some European coasts, such as The Netherlands. Adjacent barrier islands are separated from one another by tidal inlets. The inlets allow the exchange of water from lagoon to sea, and also facilitate sediment erosion, transport and deposition around the barrier island. Barrier islands are very important for defending vulnerable coastal lowlands behind them, principally by absorbing wave energy and protecting the coastline from severe storm wave conditions. Therefore, there is strong interest in studying the formation and morphodynamics of barrier island systems.

Barrier island formation

The formation of barrier islands is controversial, with three principal hypotheses that may be applicable and referred to here as:

1 **Emerged-transgressive model.** Some barrier islands may represent offshore bars that were formed during the previous glacial sea level lowstand, and that with the post-glacial rise in sea level developed vertically using accumulated sediment transported onshore during the transgression.
2 **Submerged-transgressive model.** This refers to barrier islands that may have been coastal dunes during a lower sea level, but were isolated from landward coastal lowlands through submergence and transgression by post-glacial sea level rise.
3 **Emerged-stillstand model.** The previous hypotheses consider barrier islands to represent the continued development of inherited features (offshore bars or dunes) from earlier sea level lowstands; however, this theory suggests that barrier islands have developed since post-glacial sea level rise stabilised, approximately 4,000 years ago, to produce the current sea level stillstand.

It is known that some barrier island deposits are certainly older than 4,000 years, but it is possible that all three of these models may be correct in certain situations.

Barrier island morphodynamics

Barrier islands are highly dynamic environments and susceptible to changes in wave energy, sediment supply, sea level and human interference. Mapping of barrier islands along the Gulf of Mexico (Louisiana and Mississippi) coast for more than a century has indicated the degree and types of morphodynamic changes that

Figure 2.14 Morphodynamic responses of barrier islands to changing environmental conditions.

Source: reprinted with modifications from McBride et al. (1995) with permission from Elsevier Science have occurred. McBride *et al.* (1995) have used this information to model the geomorphological response of barrier islands to natural and human variables (Figure 2.14). The model identifies eight geomorphological response types:

1 **Lateral movement.** This involves the movement of sediment along the seaward front (seaside) of a barrier island, often characterised by erosion at one end and deposition at the other end of an island. This gives the impression that the island is moving laterally along the coast.
2 **Advance.** This refers to a shore that advances seaward through progradation in response to increased sediment supply or a lowering of sea level.
3 **Dynamic equilibrium.** Refers to a shoreline that appears to be stable over long periods of time, with neither significant erosion or deposition.
4 **Retreat.** Applies to seaward-facing shores that retreat landward through the erosion and removal of sediment or from a rise in sea level.
5 **In-place narrowing.** This occurs where the seaward and landward (bayside) shores of a barrier island undergo erosion, leading to narrowing of the island, but with the core of the island remaining stationary.
6 **Landward rollover.** This often follows in-place narrowing, where the island beomes narrow enough for storm waves to overtop the island, eroding sediment from the seaward side to deposit it on the landward. Due to this rollover of sediment an island will appear to migrate landward.
7 **Breakup.** Again, this response often follows in-place narrowing, where narrowing has left the barrier island susceptible to breaching by waves to form new inlets, which may rapidly widen at the expense of barrier islands.
8 **Rotational instability.** This refers to an island that appears to be rotating, either in a clockwise or anti-(counter-)clockwise direction, in response to advance at one end and retreat at the other end of the island.

An example of barrier island geomorphological development according to the above response-type model is dramatically provided by the Isles Dernieres barrier system of the Louisiana coast in the southern USA (McBride *et al.,* 1991, 1995). It is situated on a sediment-starved coast that is experiencing a very rapid rise in sea level of 1 cm y^{-1} due to regional ground subsidence. For most of the one hundred years up to 1989, the barrier system has experienced in-place narrowing, with 11.1 m y^{-1} of erosion and retreat on the seaward shore and 1.9 m y^{-1} of erosion on the landward shore. Breaches have subsequently occurred leading to the general break-up of the system. Eroded sediment is transported out of the system or stored subtidally as shoals in the ever-widening inlets. In this way, these barrier islands have been reduced in size by 78 per cent over the study period from 3,532 to 771 ha, a rate of loss of 28.2 ha y^{-1}. Future projections based on the current rates show that this barrier system will disappear by the year 2020. The disappearance of Isle Dernieres will have severe consequences for the Louisiana coastline that it previously protected, with estuaries and fragile coastal wetlands being exposed to the full intensity of hurricane conditions. This trend may only be offset through a programme of barrier island restoration through artificial sediment nourishment.

Beaches

Beaches are perhaps the most familiar and paradoxical of all coastal environments. They are composed of loose, unconsolidated, malleable material, sand and pebbles, and yet survive the roughest storms and wave conditions that affect coastlines. Yet it is exactly this characteristic that makes beaches so endurable: sand is highly mobile and can be moved about, moulded into shapes that deal harmoniously with wave energy and so ensuring their continued existence. Hard coastal structures, whether natural (such as cliffs) or man-made (such as harbour walls), are rigid and resist wave attack, and so are eroded – beaches are seldom eroded by waves, they simply metamorphose.

Beach profiles

The way in which beaches respond to wave energy makes them excellent natural coastal defences, but the way in which they defend the coastline changes according to the level of wave energy being received (Figure 2.15). Under normal or **fairweather wave** conditions (low to moderate wave energy) beaches have quite a steep gradient (e.g. 11° or 1:8 slope) and, as we discussed earlier, this steep gradient tends to reflect the moderate wave energy back out to sea. The reason for this high beach gradient lies in the breaking of individual waves. When a wave breaks on to a beach, water travels up the beach as **swash** and after the swash has travelled as far up the beach as the energy will allow, the water returns to the sea under gravity, known as **backwash**. Under fairweather conditions, when there is a

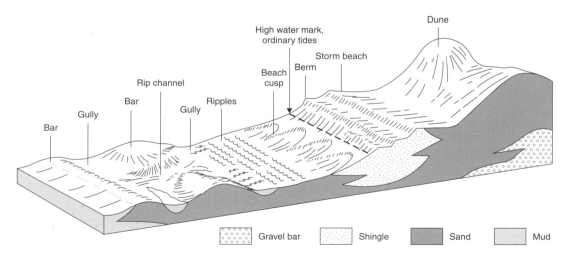

Figure 2.15 *The general geomorphology and sedimentology of a beach.*
Source: from Briggs *et al.* (1997) (Figure 17.12, p. 314)

relatively long time between consecutive waves, the backwash will often return to the sea before the next wave breaks, and so does not interfere with the swash of the subsequent wave. Because swash energy is greater than backwash energy, and if the swash of the subsequent wave is not reduced by the backwash of the previous wave, there is a greater amount of sediment transported up the beach than down it. This builds up the beach, with waves taking sediment from low down on the beach and depositing it further up. It is not surprising then that fairweather waves are often called **constructive waves**. Fairweather conditions characterise the summer, and so steep beach profiles are often called **summer profiles** (also called **swell profiles**), frequently with a prominent ridge at the back of the beach called a **berm**, which marks the limit of the swash.

Under rougher storm conditions (moderate to high energy) beaches overall have a gentler gradient (e.g. 0.5° or 1:41 slope). This gentle incline tends to dissipate wave energy because waves have a greater beach surface area over which to break. Often spilling breakers roll across these low-angled beaches for a considerable distance. Again the reason for the shape of this beach profile is due to swash/backwash interactions. This time waves arrive in relatively rapid succession at the beach, with the backwash of a previous wave returning down the beach whilst the swash of a subsequent wave is travelling up the beach. This reduces the ability of the swash to transport sediment, leading to net seaward sediment transport. In this way, sediment is taken from high up the beach and deposited seaward, thus reducing the gradient. Hence, these waves are often called **destructive waves**, and because they occur most commonly in the winter they produce a **winter profile** (also called **storm profile**). Low gradient beaches often possess an extensive sand terrace around the low tide level, which may have shore-parallel ridges or bars that separate similarly orientated depressions called **runnels** or gullies (Michel and Howa 1999). Small rip-channels breach the bars and connect up different gullies, allowing them to drain at low tide. Longshore sand bars may also occur slightly offshore.

Beach sediment

Beach sediment comes in all shapes and sizes, from rounded car-sized boulders to sand grains barely visible with the naked eye. Each sediment grain or particle, whether boulder or sand, can be described by its size, shape, and origin. Words such as boulder, cobble, pebble, sand and silt are terms used to describe **particle size** (from largest to smallest). Shape can be described in terms of a particle's roundness and sphericity. These do not mean the same thing, as a rod-shaped particle (not at all spherical) may have well-rounded ends and sides, for example. The main shape categories are:

- **discs** (flat like a coin);
- **blades** (like a matchbox);

- **rods** (long and thin like a pencil); and
- **spheres** (like a ball).

A particle's origin can be described as:

- **lithogenic** (meaning *genic* = originating from *litho* = rock, and indeed particles of pebble size or larger are commonly simply lumps of eroded rock);
- **minerogenic** (mineral grains such as quartz, produced from the break-up of rock); and
- **biogenic** (broken biological remains, such as seashells and corals).

Examining lots of particles together, as is usually the case with beach sediment, it is possible to describe the relationship of the particles to one another, otherwise known as sediment texture. Are the particles all the same size, i.e. is the sediment well sorted by size (Figure 2.16)? If the sediment consists of different sized particles, such as pebbles and sand together, then it is poorly sorted by size. It is important when discussing **sediment sorting** to state what criterion is being referred to, as sorting according to shape also occurs, and sometimes sorting by both size and shape is evident. Also, what is the composition of the sand, is it purely minerogenic, or does it also have lithogenic and biogenic particles as well, and if so in what proportions?

The character of sediment particles and texture reflect the wave environment of the beach and in turn can also influence the beach profile (Bascom 1951). High wave energy conditions are able to move nearly all particle sizes, and particle movement leads to contact with other particles, causing abrasion. Therefore, particles under these conditions are continuously being reduced in size, and yet somewhat ironically on high energy beaches, such as Chesil Beach in southern England for example, there is very little fine sediment to be seen! To understand why this is so, one must fully appreciate the dynamic relationship between waves and the beach, because the same wave energy that erodes and creates fine sediment, at the same time prohibits its deposition, keeping the finer material in suspension in the water and transporting it to quieter waters where the energy levels are low enough for it to be laid down. Therefore, all beaches may be regarded as lag deposits, consisting only of sediment that wave energy allows to be deposited, a bit like separating wheat from chaff, regardless of what is actually being produced on the beach by the same wave energy.

The influence of sediment on the beach profile is significant, and is due mainly to the ability of the sediment to let water through it, that is its **permeability**. If sediment is highly permeable and allows water through it with ease, then under any wave conditions it may be able to allow the backwash of waves to return to the sea through the beach, rather than on the beach surface. This would eliminate the backwash, and as mentioned above, allow the swash of all waves to travel up the beach unimpeded, producing net sediment transport up the beach, building up the beach and increasing its gradient. The only way the beach gradient could be reduced in such circumstances is if the beach becomes water-logged, perhaps by

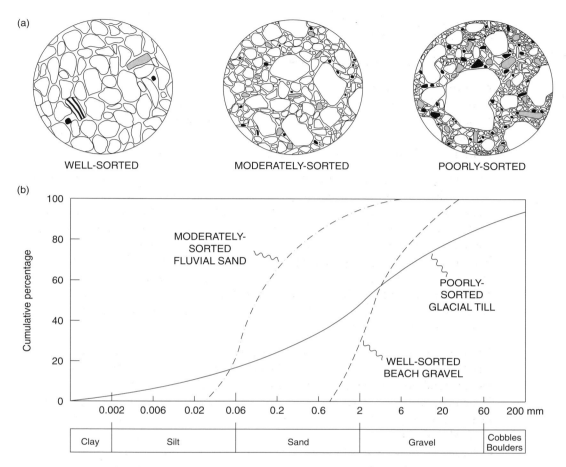

Figure 2.16 *Sediment sorting: (a) illustrates examples of well-sorted, moderately-sorted and poorly-sorted sediment grain populations; (b) cumulative percentage curves showing particle size distributions of well-sorted beach gravel alongside less well-sorted fluvial sand and glacial till for comparison.*
Source: from Briggs *et al.* (1997) (Figure 12.12, p. 201)

heavy rain or by very rapidly arriving waves. On the other hand, the backwash on beaches with low sediment permeability will return to the sea at or near the surface and may interfere with the swash, producing lower gradient beaches.

Longshore beach features

As well as variations in the beach profile discussed above, beaches often vary alongshore, but unlike beach profiles which are determined by on and offshore wave currents, alongshore variation is influenced by alongshore currents, commonly referred to as **longshore drift** (Figure 2.17a). Waves that approach a beach at an

(a)

(b)

(c)

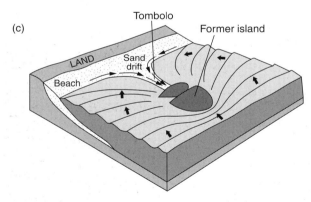

Figure 2.17 *(a) Longshore drift; (b) the formation of spits and baymouth bars as a consequence of longshore drift; and (c) formation of tombolo, an example of an equilibrium trap where two opposing longshore drift systems meet.*
Source: from Park (1997) (Figures 15.12, 15.13, 15.14, p. 428)

oblique angle may stimulate longshore drift through the oblique transport of sediment up the beach on the swash, but returning directly down the beach to the sea on the backwash under gravity, and in this zig-zag manner sediment may move alongshore until it is trapped and prevented from further transport. There are a number of natural sediment traps, of which the most important are:

- **Re-entrant traps**. These are most often embayments bounded by headlands. Sediment introduced into such a bay may undergo longshore drift within the bay itself, but is unable to escape the bay due to the high wave energy conditions affecting the headlands. Beaches that occur in such embayments are variably termed **bay-head beaches** or **pocket beaches**. If some sediment is allowed to escape around downdrift headlands, then a series of headland bound beaches may develop that expand in width downdrift as **fish-hook** or **zeta-form beaches**.
- **Salient traps**. As the name suggests, these traps project outward from the coast. **Spits** are examples that usually form where the coastline turns abruptly away from the longshore current pathway. Sediment continues to be transported and deposited linearly along the current pathway and does not follow the coastline. Spits are thus beaches that are anchored to, but largely detached from the shore and act as a repository for drifted sediment. Also, baymouth bars may form where a spit spans a bay entrance from headland to headland (Figure 2.17b).
- **Equilibrium traps**. The best examples of an equilibrium traps are

cuspate forelands. These are generally triangular features that may be formed by the convergence of two opposing longshore drift systems, with its point developing towards the minimum fetch direction. Also, local river sediment output and submarine topography may also be important in cuspate foreland development. This is certainly true for some related coastal landforms, such as **tombolos**, which are strips of sand linking offshore shoals and islands to the mainland, and may form by sediment drift and deposition in the low energy conditions of the islands' shadow zone (Figure 2.17c).

● **Deep sinks**. This collectively refers to sediment that is lost from the coast by transport into deep water, below wave-base, and so cannot be reintroduced to the coast except by sea level fall. This is often aided by the presence of submarine canyons that intersect longshore drift paths and act as conduits for coastal sediment removal.

Straight coastlines are particularly affected by longshore drift, but as long as downdrift output of sediment is matched by updrift input, the beach system will remain in equilibrium. However, a reduction of updrift input, such as natural exhaustion or commercial dredging of the sediment source, will result in narrowing of the beach and increased vulnerability of the coast to erosion and marine inundation. Remedies for sediment retention include the building of **groynes** at right-angles to the shore to trap longshore moving sediment (Plate 2.5). These are

Plate 2.5 *A groyne built to trap sand by interrupting longshore drift at Chapel St Leonards along Lincolnshire's North Sea coast (UK). Notice the well-formed ripples which indicate strong current activity along this shore.*

often effective, but in turn reduce the amount of sediment reaching sections of the beach further downdrift.

Longshore features are also present on beaches where waves break parallel to the shore, although the amount of longshore transport is limited in this context. It might be difficult to comprehend at first why any longshore drift should occur on such coastlines, however, the onshore movement of water by wave translation needs to be balanced by an offshore flow. This is manifest in **rip-currents** and ripheads, which are concentrated zones of offshore flow and are fed by the symmetrical longshore movement of water from both sides of the rip (Figure 2.18). This establishes local water (and sediment) **circulatory cells** on the beach, where water movement is generally onshore between rips, whilst at the rips it is offshore, and between the two areas the flow is alongshore. Rip-currents also occur on beaches where waves approach obliquely, but tend to be asymmetrical, being fed from the updrift direction only.

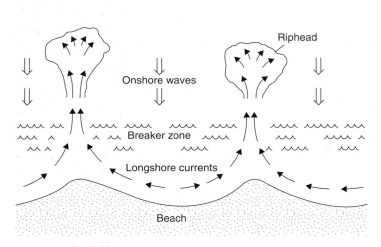

Figure 2.18 *The development of rip-currents along a swash-aligned beach.*
Source: from Briggs *et al.* (1997) (Figure 17.4, p. 305)

Commonly circulatory cells are regularly spaced along a beach and are morphologically expressed on the beach as scallop-shaped depressions known as **beach cusps** (Figure 2.15; Plate 2.6). A long-standing theory for the formation and rhythmical distribution of cusps lies in the superposition of incident waves and edge waves, created by reflection off a headland (see pp. 25–26). Where the crests of a shore-parallel incident wave and an edge wave intersect, wave-height will increase, whereas the intersection of an edge wave trough along the same incident wave crest will reduce wave-height, so producing an incident wave with variable wave-height along the length of its crest. The higher part of a breaking undulating wave crest will contain more water and have more power than the lower part of the wave crest, and so will travel further up the beach and carry more sediment with it. This extra deposition increases beach relief where the high breakers occur and subsequently act like water-sheds channelling the backwash of the high breakers into the low breaker areas. Once this process is initiated it is somewhat self-perpetuating in that erosion and offshore transport of fine sediment occur in the rips, whilst onshore transport of coarse material occurs at the built-up cusp horns. The horns also divide the swash, accentuating the circulatory cell systems. Below the water level,

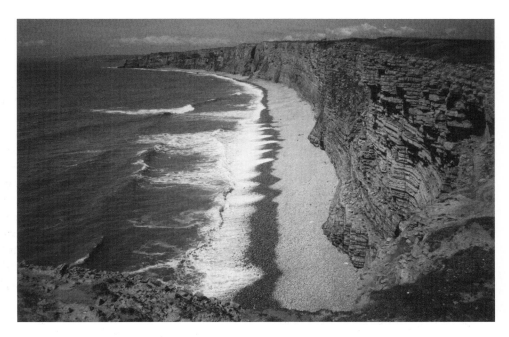

Plate 2.6 *Beach cusps formed in a gravel storm beach west of Nash Point, Glamorgan Heritage Coast (Wales, UK).*

subaqueous topography is often a mirror image of the cusps, with rip-transported sediment building up deltas that occur opposite cusp embayments, and subaqueous hollows excavated by the high breakers to provide sediment for the cusp horns.

An alternative theory states that beach cusps are formed through positive feedback between existing beach morphology and swash characteristics, and is termed the self-organisation model (Werner and Fink 1993). Irregular beach topography is enhanced by swash action, so that for example, a single upstanding positive feature on a beach would divide swash, locally creating concentrated backwash, rip-currents and embayments on either side. Embayment erosion in turn implicitly creates a horn on the other side of the embayment, opposite to the original positive feature. It is argued that the internal dynamics of the beach-swash system (hence self-organisation) transforms these irregular occurrences into regular swash circulation patterns which lead to the development of rhythmical beach cusps. The alongshore spacing of beach cusps created in this way is considered to be related to the horizontal swash excursion, which is the distance travelled by swash up the beach, so that higher swash excursions result in more widely spaced cusps (Masselink 1999).

Regardless of whether the edge wave or self-organisation model is responsible for beach cusp formation, the stability of beach cusps thereafter appears to be dependent on swash circulation. Masselink and Pattiaratchi (1998) explore the relationship between cusp prominence and spacing, swash excursion and swash circulation (Figure 2.19). They propose a model that relates three different fairweather swash circulation patterns to the further development of beach cusps:

1 **Oscillatory circulation.** This refers to swash and backwash that flow directly up and down the beach respectively, and occurs when the cusps are too widely spaced or too subdued relative to the swash excursion. The outcome of this two-dimensional circulation is that sediment from subaqueous deltas in front of cusp embayments is transported on to the beach, gradually leading to sedimentary infilling of the embayments.

2 **Horn divergent circulation.** This type of circulation occurs when swash is in equilibrium with beach cusps, so that cusp horns divide the swash, concentrating backwash in embayments. This circulation pattern maintains existing beach cusp morphology.

3 **Horn convergent circulation.** This occurs when cusps are too closely spaced or too pronounced relative to the swash excursion, so that the swash essentially swamps the cusp features, with overtopping of the horns and wave uprush in the embayments which spreads laterally to converge at the horns. This rapidly promotes erosion of the horns and deposition in the embayments.

SWASH WATER CIRCULATION	DESCRIPTION
FAIR WEATHER CONDITIONS	
(a)	OSCILLATORY • Predominantly two-dimensional flow up and down the beach • Weak flow divergence on cusp horns • Weak flow convergence in cusp embayments
(b)	HORN DIVERGENT • Swash runup is diverted from cusp horn to embayment • In the embayment, flows meet to form a concentrated backwash • Mini rips form opposite cusp embayments
(c)	HORN CONVERGENT • Swash runup enters the cusp embayment with the bore front aligned with the embayment contours • Uprush spreads laterally to the horns and forms backwash • Mini rips may form opposite cusp horns
STORM CONDITIONS	
(d)	SWEEPING • Swash runup sweeps obliquely across the beachface • Backwash follows a parabolic arc • Littoral drift is pronounced
(e)	SWASH JET • In the embayment, strong backwash retards incoming swash until it has sufficient head to overwhelm the backwash flow and rush up the beach as a narrow jet • Swash runup in the form of a swash jet fans out laterally as in (c)

Figure 2.19 *Beach cusp stability in relation to swash circulation.*
Source: reprinted with modifications from Masselink and Pattiaratchi (1998) with permission from Elsevier Science. Gerhard Masselink kindly supplied a copy of the original figure for reproduction here

Also, two swash circulation patterns are given for storm wave conditions, sweeping and swash jet circulation, but both of these are destructive.

Gravel beaches

Gravel beaches (also known as storm or coarse clastic beaches) are characterised by the dominance of large-size material (between 2–2000 mm diameter), a steep shore-face, which usually means the shoreline is of a reflective type (Figure 2.15). Unlike sandy beaches, the study of gravel beaches has been neglected, mainly because it is difficult to deploy sensitive instruments in what are quite often high-energy and destructive environments (Carter and Orford 1993).

Coarse clastic shorelines are found mainly in areas that are:

- in formerly glaciated or periglaciated areas (Plate 2.7a); or
- on tectonic coasts where high-gradient streams deliver bedload to the shore (Plate 2.7b); or
- in wave-dominated areas subject to rock cliff erosion (Plate 2.7c).

They may be classed as:

- barriers that are free-standing or fringing, with a well-defined back-slope and back-barrier depression, enclosing lagoons and wetlands, and capable of migrating inland; or

Plate 2.7 *(a) Glacial till providing a rich source of coarse sediment for gravel beach development along the southern shores of Galway Bay (Ireland)* Continued

Plate 2.7 *(b) gravel beaches along the Queensland coast north of Cairns (Australia) which are supplied with coarse sediment by high gradient streams flowing off the uplifting coastal mountains; and (c) gravel barrier at Porlock (Somerset, UK) supplied by the updrift wave erosion of cliffs (also, notice the distinct berms formed on the beach face).*

- beaches that have no well-developed wave-formed landward facing slope, and are often confined between headlands, as pocket beaches and abut cliffs, that can be further subdivided into **swash-aligned** or **drift-aligned** (i.e. perpendicular or oblique to wave approach respectively).

The form of the shoreline is crucial to the morphodynamic status of gravel shorelines. There are two basic forms:

- the first comprises a single rectilinear slope from crest to wave base (ignoring small ridges), which remains reflective under almost all wave conditions; and
- the second comprises a concave-up form, which may exhibit a marked break in slope at or around the mid- to low-tide position, which alters the morphodynamic status of the shoreline as wave height to depth and depth to wavelength ratios vary, as tide levels change. The break in slope is often mirrored by a change in sediment character.

A consequence of wave reflection is the development of edge waves, which on gravel shorelines are developed by:

- waves reflected off the shoreline and trapped by refraction (Carter 1988); and
- waves resonating between headlands, usually associated with pocket and fringing beaches.

On the shoreline, the main manifestation of edge waves is the appearance of beach cusps, and the spacing of the cusps in this context is related to edge wavelengths. Cusp morphology may exert a strong control over sediment movement and sorting. Furthermore, cusp development may dictate the pattern and position of barrier breaches and overwashing events during storms (Plate 2.8).

Permeability of the gravel shore is important in sediment transport. If the clasts are large enough, so producing sizeable interstitial pores, it may be that all swash sinks into the beach and returns to the sea through the beach. As discussed earlier, this will have the effect of minimising or eliminating backwash, so that net coarse sediment transport is landward. However, with large pore spaces, sediment decoupling may occur whereby fine sediment can be washed back through the beach to re-emerge and possibly be deposited seaward, usually at a break of slope in concave-up beaches. Here, sand may be stored in the form of a sand terrace, and this may be added to by material introduced by longshore drift.

Onshore sediment transport is concentrated within a narrow zone between breakers and the beach face and is dominated by **bedload transport**. Transport can involve either the movement of individual clasts or clast populations. Individual clasts on a flat shore are likely to move landward rapidly. Attached seaweed may aid this movement due to increasing clast buoyancy. Individual clasts tend to aggregate and as accumulations of coarse particles develop group-imposed transport controls are introduced. These controls influence sediment sorting, and there is a transition between an initial unsorted population of clasts to sorted subpopulations in terms of spatial distribution and size/shape characteristics. The net result is that gravel

Plate 2.8 *(a) A gravel barrier overwash fan at Ru Vein in the Baie d'Audierne (Brittany, France); and (b) a breach in the gravel barrier at Porlock (Somerset, UK) formed by severe storms on 28 October 1996.*

shorelines tend to become organised, in that they develop distinct cross-shore and along-shore **facies** which act to limit further transport. For example, more spherical clasts accumulate in the lower foreshore, whilst disc-shaped clasts tend to accumulate at or near the beach crest. Sorting such as this reflects wave energy, with the discs (and blades) being sorted from the more spherical clasts during high energy storms and deposited at the landward limit of storm wave activity, whilst sorting amongst the spherical clasts, into spheres and rollers, takes place in lower wave energy conditions in the lower foreshore zone (Williams and Caldwell 1988).

Box 2.5

Human-induced destabilisation of gravel beaches

Orford *et al.* (1988) document a case of a gravel barrier at Carnsore in southern Ireland where fine sediment was supplied to the shore by an outlet stream from a back-barrier lagoon. Longshore drift distributed the sand along the shoreline which created a low-angled dissipative sand terrace in front of the gravel barrier. However, sand supply was artificially stopped by a dam being constructed across the lagoon outlet. Longshore currents eventually removed the seaward low-angle slope component and the disappearance of this dissipative element led to an increase in wave energy reaching the reflective barrier. Cusp formation was initiated which provided a template for overwashing, breach formation, and general barrier degradation. Sediment was redistributed landward as washover fans.

The ecology of beach systems

Compared to many other coastal environments, sand and gravel beaches appear to be quite barren, and indeed they usually only support a living community of low abundance and diversity. Beach sediment, as we have seen, is highly mobile and regular disturbance occurs. Such an unstable substrate prohibits attachment by seaweeds that are so important in the food chain of rocky shores. However, plants do occur as microflora, such as bacteria and **diatoms**, attached to the surface of sediment grains. These are consumed by microfauna or meiofauna living in the pore spaces between grains, which in turn support macrofauna. Because of the lack of shelter on beaches and the **infaunal** occurrence of microflora and fauna, most macrofauna, such as lugworms and shellfish, have adopted a burrowing lifestyle. These are largely preyed upon by wading birds (Plate 2.9), that have evolved specialised bills of varying lengths to probe into the sand, in addition to human exploitation of shellfish. Overall, the amount of organic matter residing in beach sediment is determined by wave activity, which may directly wash it away, and indirectly impair its storage by determining grain size, as larger grains have larger pore spaces which allow leaching and oxidization to occur. Therefore, low energy beaches are often ecologically richer than high energy beaches.

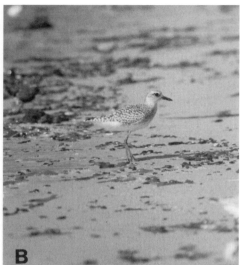

Plate 2.9 *Beach birds of southern California: during the winter months the quiet beaches around La Jolla (near San Diego) host many wintering wading birds such as (a) marbled godwit* (Limosa fedoa)*; (b) black-bellied plover* (Pluvialis squatarola), *willet* (Catoptrophorus semipalmatus), *spotted sandpiper* (Actitis hypoleucos), *black turnstone* (Arenaria melanocephala), *and sanderling* (Calidris alba) *which all exploit the beach environment in slightly different ways.*

Problems within a beach do occur, such as the drying out of the upper beach at low tide, and the frequent disturbance by turbulent surf activity of the lower beach has led to ecological zonation, but it is not as clear and well defined as in rocky shores. At and above the high tide limit, plants rapidly colonise beach sediment, perhaps leading to dune formation on sandy beaches, stabilising gravel on coarser beaches, and generally increasing the abundance and diversity of species.

Coastal sand dune systems

Coastal sand dunes occur landward of the shoreline, usually above the high tide level, and are often perceived to mark the landward limit of marine influence on the coast. The inclusion of sand dunes in a chapter on wave-dominated coasts at first sight may seem slightly inappropriate as dunes are mainly formed by aeolian (wind) processes and not by waves. However, many coastal dune systems are genetically related to sandy beaches that occur seaward of them, and these beaches are often characteristic of wave-dominated coasts as discussed earlier. They are extremely important coastal landforms as they often act as a coastal defence, protecting coastal lowlands from marine inundation. It is for this reason that coastal dunes are extensively studied from geomorphological, ecological and management perspectives.

A number of conditions are required for the formation of coastal sand dunes, including:

- An area landward of the beach that is able to accommodate blown sand, usually where cliffs are lacking.
- A strong onshore wind for transporting sand from its source on a beach to the dunes. For this reason, dunes are particularly common along coasts frequently affected by storm conditions, such as northwest Europe and northwest USA. Indeed, the height attained by a dune is determined by wind velocity, so that higher dunes occur in the stormiest regions.
- Suitably sized sand and an abundant supply of it. Without sand capable of transport by aeolian processes dunes cannot form. Also, for dune maintenance and continued development a further supply of suitable sand into the system is needed. Hence, larger and well-developed dunes are commonly situated close to sediment sources, such as river outlets delivering sediment from erosive catchments to the coast.
- Vegetation to colonise and stabilise blown sand (Figure 2.20a). Unvegetated dunes of the kind normally seen in arid deserts can occasionally occur at the coast, either where deserts are adjacent to the coast (e.g. Namibia, Africa) or where the rate of sand movement is high and prevents the establishment of vegetation (e.g. some dunes along the Washington and Oregon coast, USA). These unvegetated dunes are called **free dunes** and are sensitive to wind direction, often orientating themselves at right angles to the prevailing wind. However, most coastal dunes are colonised to varying degrees by vegetation and are called **impeded dunes**. Vegetation has a stabilising effect and to a large extent prevents sand loss and dune migration inland. The orientation of impeded dunes is more aligned with the source beach rather than wind direction.
- A low gradient of the source beach coupled with a large tidal range. The combination of these two parameters means that large expanses of beach sand are exposed at low tide, increasing the time available for sand to dry and for the wind to transport it landward. Drying out of the beach is particularly important and is one of the reasons why dunes are generally absent along tropical coasts, where high atmospheric humidity suppresses drying.

Aeolian sand transport and deposition

The initial movement of sand by wind occurs when a critical wind velocity is attained relative to a given sediment particle's size, and this is termed the **fluid threshold velocity** (Figure 2.20b). The relationship is generally positive with increasingly higher wind threshold velocities required to move coarsening particles. However, this relationship is reversed for very fine particles (clay and silt), which

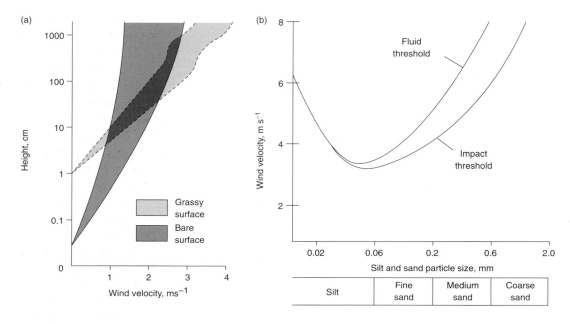

Figure 2.20 *(a) A comparison of wind velocity over bare and grass (10 cm tall) surfaces clearly indicating that vegetated surfaces prohibit deflation; (b) fluid and impact threshold velocities for different particle sizes.*
Source: from Briggs *et al.* (1997) (Figure 16.1, p. 293)

resist movement due to a high degree of cohesion between grains. Once initial sand movement has been established, transport may proceed as **surface creep**, where particles roll along the surface. Alternatively, if wind velocity is sufficient, sand may be entrained into the airflow and transported in suspension. A combination of these two processes produces the commonest mode of aeolian sand transport, that of **saltation**. This process involves particles that jump or hop which, after initially being stimulated at the fluid threshold velocity, are entrained into the airflow for a short distance before falling back to the surface under gravity. On landing, saltating particles impact with other grains and transfer kinetic energy, therefore reducing the threshold velocity required by the other particles to move. This lowered threshold is known as the **impact threshold velocity** and means that once sand transport is initiated it can be maintained by lower wind velocities (Figure 2.20b).

The deposition of sand requires a reduction in wind velocity. On a beach this reduction occurs most commonly in the lee of obstacles, such as debris (seaweed, shells, pebbles, drift wood, litter, etc.) found along a strandline at the high tide level. Sand accumulations quickly develop a streamlined duneform in response to the wind conditions, which is characterised by a gently sloping upwind stoss side and higher gradient downwind lee side (Figure 2.21). The internal sedimentary structure of dunes commonly exhibit co-occurring low and high angled bedding planes, known as **cross-bedding**, that represent old stoss and lee surfaces respectively.

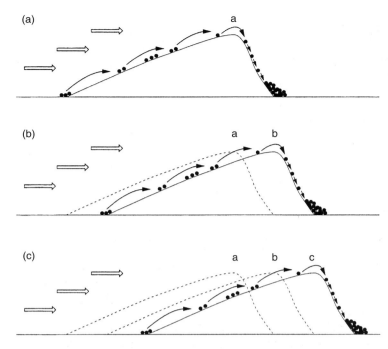

Dunes forming around debris build up to form low and unvegetated **shadow dunes**. Pioneer vegetation may subsequently colonise these mounds to form **embryo dunes** (Plate 2.10). The presence of vegetation traps further sand and allows the dune to develop substantially above and beyond the original piece of debris. In this way neighbouring embryo dunes can coalesce to form a dune ridge coincident with the initial strandline position, so forming **foredunes** running along the back of the beach.

Figure 2.21 *The formation (a) and migration (b and c) of sand dunes, also indicating the development of internal cross-bedding, where stoss and lee-sides are preserved as oblique bedding planes.*
Source: from Park (1997) (Figure 13.8, p. 366)

Dune system morphodynamics

A single foredune ridge is the basic geomorphological requirement for the establishment of a coastal dune system. However, the morphodynamics of a dune system is largely dependent upon the nature of further sediment supply from the source beach (Figure 2.22):

● If sediment supply is low, then sand blown inland from the foredune may not be replaced, leading to a decrease in foredune volume and rendering it vulnerable to erosion, particularly cliffing or scarping of the foredune front by waves during storms and fragmentation through the development of **blow-outs** along the fore-dune ridge. Blow-outs represent the localised removal of sand by the wind, a process called **deflation**, which create relatively small hollows. Blow-out formation is often stimulated through disturbance of the dune by human activity (walking, riding, etc.), animal activity (e.g. rabbit burrows, grazing), or storm wave activity, which are all capable of undermining fragile dune vegetation, leading to sand liberation. Sand deflated from blow-outs is often re-deposited downwind, where vegetation remains intact, in the form of crescentic **parabolic dunes**.

Plate 2.10 *Embryo dunes forming in discrete mounds at Barneville (Normandy, France).*

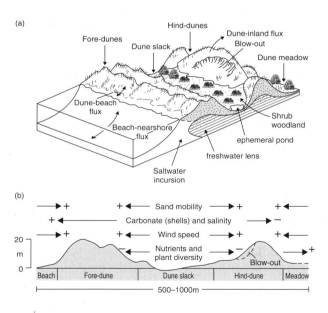

Figure 2.22 *(a) The geomorphology of a coastal dune system; and (b) the distribution and influence of some environmental parameters affecting coastal dunes (+ = increase, − = decrease).*
Source: from Briggs *et al.* (1997) (Figure 16.6, p. 300)

- Foredune morphology will be maintained if the loss of sand from the system is matched by new supply from the beach, so that there is no net change in sand storage within the dune.
- Where sediment supply from the beach is greater than is being lost from the dune (i.e. a net sediment gain), and where the coastal slope is relatively gentle, then dune systems may prograde seaward through the formation of new foredune ridges. In this way, a whole series of dune ridges may be formed, but only the most seaward is a foredune, whilst landward dune ridges are referred to as second, third and fourth hind-dune ridges, and so on. Depressions formed

between the ridges are called **dune slacks**. These are often damp (**wet slacks**) where the surface of the slack intersects the water table, which generally limits deflation due to the presence of wet sand on the slack surface. However, water table fluctuations can sometimes lead to flooding in the winter and drying in the summer. The morphology of dune ridges and slacks is maintained by an undulating air flow or 'wind-waves' over the dune surface (Pethick 1984). Continued progradation increasingly distances older dune ridges from the source beach so that eventually they may receive very little new sand. This leaves them vulnerable to blow-out formation which at its extreme may transform the initial shore-parallel or **primary dune** ridge into a number of shore-normal **secondary dune** ridges that comprise sand reworked within the dune system.

Coastal dune ecology and management

Plants are of vital importance to the formation and stability of a coastal dune system (Figure 2.20a). However, dunes provide a very harsh habitat, with a highly mobile sand substrate, very little water retention (except in wet slacks) coupled with the drying effect of the wind, high salt input, extreme ground temperatures and minimal nutrient and organic matter content of the sand. Even so, at the high tide level salt-tolerant pioneer plants colonise shadow dunes to form embryo dunes, and once these pioneers become established other plants are able to take hold, especially marram grass (*Ammophila arenaria*), leading to the ultimate development of foredunes. Foredunes are characterised by such marram grass or yellow dune communities, so named because of the great amount of bare sand that is visible on the dune surface. This community is largely **xerophytic**, comprising plants that can survive conditions with minimal water, relying on their extensive rhizomous root system and thick shiny leaves to obtain and conserve water respectively. Landward this gives way to grassland/heathland or grey dune communities and eventually scrub/woodland or climax communities. Wet slack communities are similar to freshwater bogs, marshes or wet meadows, and are often characterised by quite high floral and faunal diversity (Cremona 1988).

Dune systems are extremely sensitive and disturbance that leads to destruction of vegetation more often than not results in deflation and blow-out formation. Blow-outs can seriously affect the integrity of the dune system and its coastal defence capability. Commonly, blow-outs in foredunes act as a template for breaching by storm waves, which may lead to flooding of lowlands landward of the dunes. Also, sand blown inland has in the past engulfed entire villages. Where sand supply is abundant, blow-outs are only temporary and the dune system is not necessarily at risk. Similarly, dune scarps commonly develop in the winter as storms erode the foredunes, removing sand to subtidal longshore bars, but then this sand is returned under the summer constructive regime. However, many dune systems have very limited new sand input, upsetting this state of dynamic equilibrium and so are very

Plate 2.11 *Dune scarp formed during winter storms and persisting into the summer, indicating limited new sand supply, at Genets in the Baie de Mont St Michel (Normandy, France).*

susceptible to sand loss (Plate 2.11). Therefore, the management of coastal sand dunes is a major concern for many coastal protection agencies throughout the world.

One of the main sources of dunes' disturbance comes from their use as a recreational resource. A number of measures can be taken to minimise the effect of visitor pressure. Generally visitors to dunes are simply seeking access to a beach, and it is while they are making their way there that damage occurs. Therefore, only a few large car parks should be provided per dune system, rather than many small car parks, as this reduces the number of access points to the beach, so minimising potential foredune fragmentation. Walkways from these car parks should be managed to avoid pathway incision and the undermining of adjacent vegetation. This may be achieved through building boardwalks or other similar hardstanding structures, and also by prohibiting visitors wandering off the walkways. Other activities, such as on-dune driving, horse riding and camping, should also be prohibited on vulnerable dunes.

Where disturbance has already occurred, some form of remedial action is required. First, areas of bare sand within a blow-out, for instance, should be stabilised either by planting marram or by the use of artificial material, such as nets, collectively known as geotextiles. Second, sand deposition may need to be encouraged, mainly through the use of artificial sand traps, commonly in the form of picket fences that allow wind to pass through them, but reduce wind speed enough for deposition to occur. Such fences also have the additional benefit of restricting access to these vulnerable areas.

Summary

- Waves are created by solar, seismic or cosmic energy, and in deep water only transfer energy, and not matter.

- In shallow water waves transmit energy and matter, facilitating erosion of the coast, and transportation and deposition of sediment.

- Geomorphology of wave-dominated coastal systems is determined by the input and output of wave energy and material (i.e. sediment). Coasts with high energy and low sediment input typically form erosional coastlines, whilst high sediment input systems tend to be depositional.

- The landforms and ecosystems of wave-dominated coasts are sensitive to interference from human activity.

Discussion questions

1 What is the geomorphological significance of nearshore wave modification processes?

2 Given that high wave energy promotes erosion, why do high wave energy beaches generally lack fine sediment?

3 To what extent are wave-dominated coastal systems influenced by human activity?

Further reading

See also

Geological and tectonic coastal classifications, Chapter 1.
The geomorphological significance of tidal range, Chapter 3.
Coastal responses to sea level change, Chapter 5.
Approaches to coastal zone management, Chapter 6.

General further reading

Waves, Tides and Shallow Water Processes. Open University. 1989. Pergamon Press, Oxford, 187pp. A thorough text focusing on the operation of coastal processes, including waves.

Tsunamis in the World. S. Tinti (ed.). 1992. Kluwer Academic, Dordrecht, 228pp.

Geomorphology of Rocky Coasts. T. Sunamura. 1992. Wiley, Chichester, 302pp. An excellent and technical account of erosional processes and resultant landforms at the coast.

Coral Reef Geomorphology. A. Guilcher. 1988. Wiley, Chichester, 228pp. An introduction to the physiology of coral reef systems throughout the world.

Islands at the edge. J. Ackerman. 1997. *National Geographic*, 192(2), 2–31. A colourful and informative account of the physical, ecological and human aspects of barrier islands.

Beaches: Form and Process. J. Hardisty. 1990. Unwin Hyman, London, 324pp. An advanced-level text introducing beach and nearshore dynamics.

Coastal Dunes: Form and Process. K. F. Nordstrom, N. Psuty and B. Carter (eds). 1990. Wiley, Chichester, 392pp. A collection of papers dealing with the development and geomorphology of coastal dunes with good case study material.

3 Tidally-dominated coastal systems

For most coasts, the importance of tides lies in the way they help to determine the areal extent of the coastline that is affected by marine processes. Only in certain coastal environments, such as estuaries, do tides themselves and the currents they produce become a powerful agent of erosion and deposition. This chapter covers:

- the generation of tides by the moon and the sun and the resulting tidal levels
- the importance of tidal range for coastlines and coastal processes, such as tidal current activity
- the hydrodynamics, geomorphology, sedimentology and ecology of estuarine systems, including associated coastal wetlands, such as salt marshes, mangroves and sabkas
- the impact of human activity in tidal coastal systems

Introduction

Tides are a natural phenomenon that most people are aware of to a degree. They represent an everyday part of living at the coast and, indeed, many coastal activities (e.g. sailing, fishing, beachcombing, etc.) are strongly influenced by the state of the tide. But tides are cyclical and very predictable, and because of this people living at the coast must have accepted at a very early stage the control exerted by the tide on their lives and the pace at which they live.

All coasts are influenced to some extent by tides, but only a few types of coastal environments can be said to be tide-dominated. Amongst those tide-dominated environments are estuaries, and these figure prominently in the settlement geography of the world. This is because estuaries are obvious trade routes and more often than not offer sheltered harbourage to sea-going trade vessels, and because estuarine-fringing wetlands are usually flat surfaces which are ideal for urbanisation once they have been reclaimed and drained. Many major global urban centres are located in estuaries, such as London in the Thames estuary, so that although tide-dominated coastal environments may be regarded as uncommon, they support a disproportionate level of the global population. Thus estuaries are particularly affected by urban effluent discharge and other sources of pollution. This chapter introduces tides as they apply to all coasts, but then goes on to discuss in some detail the tide-dominated coastal systems of estuaries and their associated tidal wetlands, the salt marshes, mangroves and sabkas.

Tides and their generation

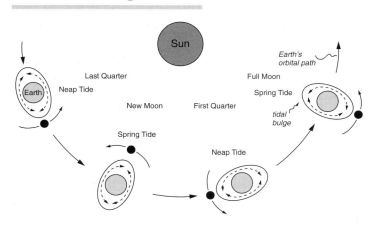

Figure 3.1 *The formation of the tidal bulge with regard to the relative position of the earth, moon and sun.*
Source: from Briggs *et al.* (1997) (Figure 4.11, p. 60)

Tides are in fact waves with extremely long wavelengths. They result from the gravitational attraction of the sun and the moon, and although the sun is a much larger body, the moon is more influential because it is nearer the earth. As the sun and moon pass overhead they attract the surface ocean creating a **tidal bulge** (Figure 3.1). Due to centrifugal forces, a second tidal bulge also occurs on the opposite side of the globe. When one of these 'tidal waves' meets the coast, the crest produces a high tide whilst the trough produces a low tide. The magnitude of tides produced in this way changes according to the relative positions of the earth, moon and sun during their orbital cycles. When the sun and moon are aligned with respect to the earth, when the moon is new or full, their component gravitational effects are combined to produce higher than average high tides known as **spring tides**. Conversely, when the sun and moon are at right angles with respect to the earth the gravitational attraction is dispersed with the tidal bulge and the resulting high tides are lower than average. This is known as a **neap tide**. Similarly, the low tide levels associated with spring and neap tides are respectively lower and higher than average. The spring–neap tidal cycle is approximately fourteen days, so that it takes two weeks to go from spring to neap to spring tide. The largest of the spring tides occur at the vernal (spring) and autumnal equinoxes, when the sun crosses the equator.

Along most coasts, including nearly all open Atlantic coastlines, there are two high tides and two low tides every day, the **semi-diurnal tides**. High and low water are separated here by approximately 6 hours and 13 minutes, so that high and low water occur slightly later each day. However, some coasts experience only one high and one low tide a day due to local factors, these are **diurnal tides**. For example, diurnal tides occur at Do-Son in Vietnam and New Orleans in the USA. And along a few coasts the two are mixed, where the daily tides are dominated by one extreme high and low tide, but smaller secondary high and low tides do occur. This **mixed tide** is typified by tides along the western seaboard of the USA, such as at San Francisco and Los Angeles (Figures 3.2 and 3.3).

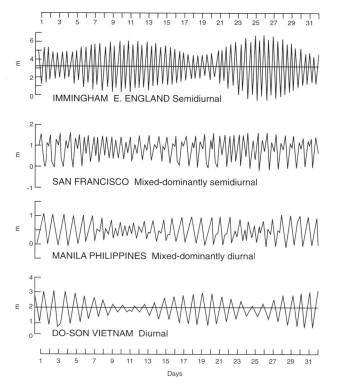

Figure 3.2 *Examples of semi-diurnal, mixed and diurnal tidal cycles.*
Source: from Briggs *et al.* (1997) (Figure 4.12, p. 60)

It is clear from the above discussion that the regularity of orbital cycles produces predictable variations in tides at a number of different timescales from daily to decadel, and hence **tidal cycles**. Indeed, an important tidal cycle that can affect the establishment of tidal levels (e.g. mean sea level) takes 18.6 years to come full circle.

The spinning of the earth beneath the tidal bulge makes the bulge appear to travel uninterrupted around the globe and on a landless earth it would do this. However, in reality the continents get in the way of the tidal bulge and as with all intercepted waves reflection occurs. In this way tidal bulges are pulled by the moon and sun westward across oceans, and upon meeting the eastern side of continents travel eastward back across the oceans as reflected waves just in time to meet the moon and sun coming round again. This simple to and fro scenario does not actually occur, however, due to the **Coriolis effect** which under the influence of the earth's rotation causes travelling water, winds and other moving objects to appear to curve to the right (clockwise) in the northern hemisphere and to the left (anticlockwise) in the southern hemisphere – just look at water going down the plughole, it swirls clockwise and anticlockwise in the northern and southern hemispheres respectively. This means that any given tidal wave travels in a circular manner known as **amphidromic motion**, and every ocean basin possesses an **amphidromic system**. An **amphidromic point** marks the centre point of an amphidromic system, and the positioning of these points in an ocean basin depends on the geometry of the ocean basin, including coastal configuration and bathymetry, and they can even occur inland as degenerate or degraded amphidromic points.

Figure 3.4 shows, for example, the amphidromic systems around the British Isles. The radiating lines are co-tidal lines which show the position of the tidal wave in hours from the beginning of a tidal cycle. Therefore, it is clear that high water travels around the amphidromic point, so that low water occurs on the opposite side of the system to high water at any given time. This means that the water level at the

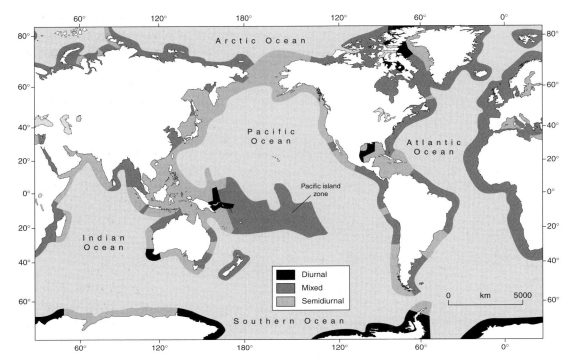

Figure 3.3 *The distribution semi-diurnal, mixed and diurnal tidal cycles around the global coastline.*
Source: from Briggs *et al.* (1997) (Figure 17.16, p. 318)

amphidromic point does not change, remaining unaffected by the tidal wave. The further one travels away from the amphidromic point along a co-tidal line, the more extreme will be the tidal excursion from mean sea level, with increasingly higher high tides and lower low tides.

Tide and datum levels

Tidal cycles can be predicted and used to produce tide tables for particular locations, although these predictions are supplemented to a high degree by tidal measurements made by tide gauges. Tide gauges allow tidal levels to be established, some of which are employed as survey data for making maps and charts. The most important of these levels are:

● *high and low water (HW and LW)* – the maximum and minimum tide attained during any one tidal cycle respectively;
● *mean high and low water springs (MHWS and MLWS)* – the average spring high and low water levels respectively over a period of time;
● *mean high and low water neaps (MHWN and MLWN)* – the average neap high and low water levels respectively over a period of time;

Figure 3.4 *The amphidromic systems in the seas around the British Isles.*

Source: from Briggs *et al.* (1997) (Figure 4.13, p. 61)

- *mean high and low water (MHW and MLW)* – the average of all high and low water levels respectively over a period of time;
- *mean higher and lower high water (MHHW and MLHW)* – the average of the higher and lower high water levels respectively that occur in each pair of high waters in a tidal day (approximately 24 hours and 50 minutes) over a period of time;
- *mean higher and lower low water (MHLW and MLLW)* – the average of the higher and lower low water levels respectively that occur in each pair of low waters in a tidal day over a period of time;
- *mean sea level (MSL)* – average of water levels observed each hour over a period of at least a year,

but preferably over a period of nineteen years, so as to encompass the 18.6 year tidal cycle;

- *mean tide level (MTL)* – average of all high and low water levels recorded each day – this will usually differ only slightly from mean sea level; and
- *highest and lowest astronomical tide (HAT and LAT)* – the highest and lowest water level respectively that is predicted to occur under any combination of astronomical conditions.

For land maps, either mean sea level or mean tide level are most frequently used as datums, depending on the detail of the tide gauge record available. In the United Kingdom, for example, mean sea level determined from six years (1915–1921) of continuous tidal records at Newlyn in Cornwall is used by the Ordnance Survey as their datum, and called Ordnance Datum (Newlyn), although present UK mean sea level is approximately 0.1 m higher than recorded in the period 1915–1921. Sea charts, however, use local low water or lowest astronomical tide as Chart Datum, so that below this level there should always be some water under the keel. Tide tables give tidal heights relative to Chart Datum so that mariners add predicted tidal heights to Chart Datum to obtain water depth at a given time. Because Chart Datum is based on local low water levels, it does not represent a horizontal plane, so that the height difference between Chart Datum and mean sea level varies from place to place.

Meteorological effects and storm surges

Observed tidal levels may be significantly different from predicted levels. When this occurs it is usually attributable to meteorological effects. The ocean surface essentially acts like a barometer, with a rise or fall of the sea surface of 1 cm for every millibar of change in atmospheric pressure. For example, a high pressure atmospheric system 50 mb above average would lower sea level by 50 cm, whereas low pressure of a similar magnitude would allow sea level to rise by 50 cm. When low atmospheric pressure and predicted high tides co-occur, the tide level will be higher than anticipated and quite often the flooding of coastal lowlands takes place. This can happen on a very damaging scale when the low pressure is combined with storm conditions, with very high onshore winds. Where these storm winds blow water into semi-enclosed seas at the time of high tide, the water level can be piled up to several metres higher than predicted, resulting in a **storm surge** (Figure 3.5). Extensive coastal flooding can occur in association with storm surges and this is most devastating along low-lying coasts where its effects can extend many kilometres inland. Storm surges can be geomorphologically significant in that they may overtop or breach relatively low coastal features, such as dunes, and they often lead to wave attack at higher levels than attained by normal wave conditions.

Cyclones commonly produce storm surges along tropical coastlines (Box 3.1), but elsewhere they are relatively rare, although the North Sea, due to its configuration, is regularly threatened and suffered a major event in 1953 which caused the loss of thousands of lives in eastern England and The Netherlands. This event stimulated the setting up a storm warning service for eastern England and the construction of the Thames Barrage, to protect London from a similar disaster in the future.

Tidal range

The height difference between high water and low water during the tidal cycle is known as the **tidal range**. As discussed earlier, tidal range increases with distance from an amphidromic point, so that a coastline located near an amphidromic point experiences a small tidal range, whilst a coast on the periphery of an amphidromic system will experience a much greater tidal range. In addition, a number of other factors contribute to the great variety of tidal ranges experienced on the world's coasts. These include:

Figure 3.5 *A storm surge in the Irish Sea (British Isles) resulting from a deep depression in January 1975.*
Source: from Briggs *et al.* (1997) (Figure 17.5, p. 305)

- **Bathymetry** – because of the enormous wavelength of the tidal wave it can be considered everywhere as a shallow water wave. Therefore, it can undergo refraction like all waves and become focused on particular stretches of coastline, where tidal energy, height and range are increased.
- **Width of continental shelf** – the very shallow waters encountered by a tidal wave on a continental shelf reduce celerity and increase the wave height. The slowing of the front of an approaching tidal wave allows the back of the wave to catch up, so increasing wave height further. Therefore, wider continental shelves allow more time for the broad tidal wave crest to concentrate into a narrower but higher wave, increasing tidal height at the coast.
- **Coastal configuration** – tidal waves entering coastlines which are restricted in some way, such as embayments, gulfs and estuaries, will become compressed as they progress, increasing tidal height and range. This is most marked in funnel-shaped estuaries where the estuary width decreases incrementally upstream, such as the Severn Estuary in southwest Britain where the tidal range is in excess of 14 m. Conversely, open ocean coasts, often coupled with a narrow continental shelf, tend to reflect the tidal wave, resulting in minimal tidal ranges. Nichols and Biggs (1985) examined the influence of configuration on tidal range in estuaries. They state that variations in tidal range along an estuary are determined by the relationship between the upstream

---

Box 3.1

Storm surge – the Indian 'Super cyclone' of 1999

The Meteorological Office in the UK report that in October 1999 a deep cyclone, labelled 05B, developed in the Bay of Bengal and tracked towards the eastern coast of India. It made landfall in the Indian state of Orissa in the morning of 29 October, where winds peaked at 255 km per hour. Ten million people are thought to have been affected, one million made homeless, and the death toll as of 10 November was 7,500, but was expected to rise. These deaths were not due, however, to the high winds, but to flooding from heavy rain and a severe storm surge. Some unconfirmed reports suggest that the storm surge was as much as 6 metres high and inundated the coast to up to 14.5 km inland.

---

convergence of the estuary sides and the friction created between the tidal waters and the estuary bed (roughly equivalent to the surface area of the estuary), because increased friction will diminish the tidal range. For estuaries where the effect of convergence is greater than friction, tidal range will increase up estuary, producing a **hypersynchronous estuary**, and in these estuaries a **tidal bore** wave may be produced heralding the incoming tide. Where convergence equals friction then tidal range will be uniform along the length of the so-called **synchronous estuary**; and for estuaries where the effect of convergence is less than friction, then tidal range will decrease up the estuary to produce a **hyposynchronous estuary** (Dyer 1997).

Coasts can be classified according to their tidal range (Davies 1964), with three categories being recognised (Figure 3.6):

1 **Microtidal** coasts are those that experience tidal ranges of less than 2 m and are characteristic of open ocean coasts, such as the eastern seaboard of Australia and the majority of the Atlantic African coastline, for example.
2 **Mesotidal** coasts possess tidal ranges between 2–4 m according to Davies (1964). Some authors, however, consider mesotidal coasts to have tidal ranges between 2–6 m (e.g. Briggs *et al.* 1997). However, the majority of texts adopt Davies' definition (e.g. Pethick 1984; Carter 1988; Summerfield 1991; Viles and Spencer 1995; French 1997). Examples of mesotidal coasts include much of the Malaysian and Indonesian coastline, and along the eastern seaboard of Africa.
3 **Macrotidal** coasts experience tidal ranges in excess of 4 m according to Davies (1964) and most other authors, but defined as greater than 6 m by Briggs *et al.* (1997), a range considered by some authors to indicate hypertidal conditions. Macrotidal coasts occur where the continental shelf is wide allowing the shoaling tidal wave to increase in height, and where the coastal configuration amplifies the tidal height. Examples include most of the northwest European coastal seas (e.g. Celtic Sea (including the Severn Estuary), the North Sea, the English

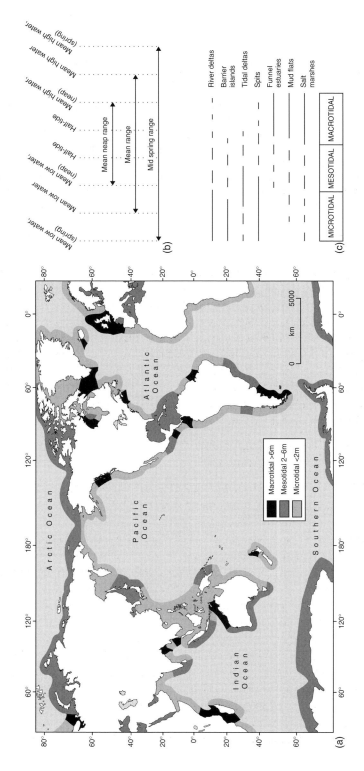

Figure 3.6 *(a) The distribution of tidal ranges around the global coastline; (b) the variation of tidal range during monthly tidal cycles; and (c) coastal geomorphological features associated with the various tidal range categories.*

Source: from Briggs *et al.* (1997) (Figure 17.17, p. 318)

Channel, and the Bay of Biscay), and parts of northeastern North America (e.g. Hudson Bay and the Bay of Fundy).

Macrotidal coasts are considered to be tide-dominated, in that most erosional, transport and depositional processes operating here are driven by tidal forces. Mesotidal coasts are considered mixed with wave and tide processes being equally important. Microtidal coasts however, are wave dominated, and many of the coastal systems discussed in Chapter 2 are of this type. Overall, there is a close relationship between tidal range and the type of coastal landforms encountered along any given coastline, and therefore, an appreciation of tidal range is essential in understanding coastal systems' diversity. For example, estuaries with their component tidal flats (whether sand or mud flats), salt marshes or mangroves, and characteristic ecosystems are typical of coasts with high tidal ranges, whereas at the other end of the spectrum barrier islands (as discussed on pp. 45–49) are commonest along microtidal coasts.

Tidal range is particularly important for coastal geomorphology because it influences the operation of physical processes (Figure 3.7). There are a number of reasons why such importance is attached to tidal range:

● Tidal range and the gradient of a coast together determine the horizontal extent of the **intertidal zone**, which is the area lying between high and low water. High-gradient, microtidal coasts have the smallest intertidal zones, whilst low-gradient, macrotidal coasts have extremely extensive intertidal areas. Ecological diversity in intertidal zones is often greater and more complex on macrotidal

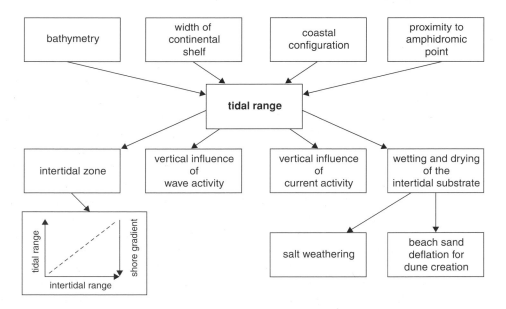

Figure 3.7 *The geomorphological signficance of tidal range.*

coasts, with salt marshes occurring high and unvegetated tidal flats low within the intertidal zone.

- It determines the extent of the vertical distance over which coastal processes operate, especially wave activity. On microtidal coasts wave breaking is concentrated within a very narrow vertical zone throughout the tidal cycle, and it is under these conditions that well-defined erosional features such as wave-cut notches are preferentially formed. However, wave energy on macrotidal coasts can be distributed over many metres throughout the tidal cycle, so that its erosional capacity is relatively diminished, but it does mean that wave activity influences a wider area. The same is also true for the operation of tidal currents (see pp. 83–86), with a greater tidal range subjecting a wider area to their activity.

- The periodic rise and fall of the tides causes wetting and drying of the substrate within the intertidal zone. Generally, the greater the tidal range, the more substrate is exposed or submerged at different states of the tide. This is important for a number of processes, including salt weathering, where seawater invading hard crystalline or laminated rocks submerged at high tide is evaporated when exposed at low tide. Salt crystals grow within minute voids in the rocks producing stresses which weaken and ultimately lead to the disintegration of the rock. This is particularly prevalent along tropical coasts, where evaporation and salt crystal growth can follow rapidly after exposure, and where especially susceptible rock types, such as granite, are found at the coast. Also, as discussed on pp. 64–71, sand dunes are more likely to develop on coasts with a relatively high tidal range, as this provides a wider expanse of beach sand for drying and subsequent landward transport by aeolian processes.

Tidal Currents

As water rises and falls with the tides it produces **tidal currents**. A rising tide that floods the intertidal zone is known as a **flood tide**, whilst a falling tide is the **ebb tide**. The significance of tidal currents lies in their ability to entrain and transport sediment. Maximum tidal current velocity is achieved at the flood and ebb tide mid-points, then at high and low water current velocity decreases to zero (slackwater) before reversing. Therefore, maximum sediment transport occurs at the mid-points in the flood and ebb tides (i.e. midway between high and low water and vice versa), whilst sediment deposition predominantly occurs around slackwater (i.e. at high and low water) (Figure 3.8a). Critical tidal current velocity thresholds exist for transporting different particle sizes. Transport may occur at and above the threshold velocity for a given particle size, but deposition occurs where velocity is less than the threshold. Figure 3.8b illustrates this with reference to mud and sand-sized particles and shows the relationship between current velocity and sediment size distribution within the intertidal zone, with mud characterising the low-energy low and high intertidal areas, whilst sand shoals occur in the high-energy mid-intertidal

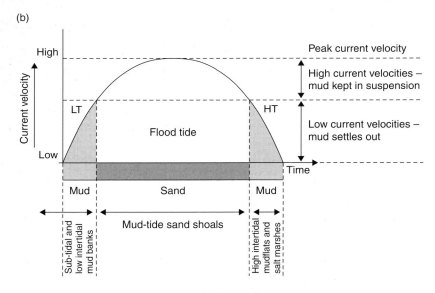

Figure 3.8 *Tidal current activity and sediment deposition: (a) changes in current velocity and direction through a tidal cycle; and (b) the relationship between flood tide current velocity and the distribution of different sediment types in the intertidal zone (HT, high tide; LT, low tide).*
Source: from French (1997) (Figure 2.10, p. 45)

zone. Overall much sediment is transported in, out and within the intertidal zone during every ebb–flood tidal cycle.

Quite often a **sediment couplet** may be deposited (Plate 3.1), where sand deposited during tidal current deceleration is overlain by mud deposited at slackwater. If an additional couplet is deposited during each cycle then stacks of couplets may be preserved as **tidal rhythmites**. These have been used to reconstruct past tidal

Plate 3.1 *A salt marsh cliff in the Baie de Mont St Michel (Normandy, France) showing distinct sediment couplets. Note coin for scale (c. 2cm diameter) near top of the cliff.*

regimes, with the thickness of sand layers indicating spring to neap tide variations, with the thicker sand layers representing the higher current velocity associated with spring tides. Indeed, if neap tide current velocities fail to exceed the sand transport threshold then the neap couplets will be dominated almost exclusively by mud (Reading and Collinson 1996).

In many instances, due to factors similar to those discussed on pp. 74–76, the ebb–flood tidal cycle is asymmetrical, with either the flood or ebb tide taking longer than the other. For example, a twelve-hour symmetrical tide consists of 5.5 hours per flood and ebb tide element, with one hour slackwater at high tide; however, an asymmetrical tide might comprise a flood tide lasting 3.5 hours and an ebb of 7.5 hours (French 1997), with one hour at high water. This is of great significance for tidal geomorphology because a fixed volume of water or **tidal prism** must come in and go out each ebb–flood tidal cycle, which means that during the shortest tidal element, which in the above example is the flood tide, the tide must flow faster. The faster tidal velocity associated with the flood tide, in the above example, means that more and coarser sediment can be transported during the flood, which becomes the **dominant current**. The ebb tide, however, becomes the **subordinate current**. A tidal coast may be referred to as flood- or ebb-dominated, depending upon which tidal element provides the dominant current. Extending this to sediment transport suggests that there is a net landward transport of sediment in flood-dominated

Box 3.2

Tidal sedimentary bedforms

The symmetry of the ebb–flood tidal cycle influences the types of sedimentary bedforms that occur within the intertidal zone. Sediment bedforms in tidal environments can be divided into two categories:

1 **Parallel orientated bedforms** are those with their long-axis orientated in the same direction as the tidal current flow and include furrows, gravel waves and sand ribbons.
2 **Perpendicular orientated bedforms** are those with their long axis at right angles to the tidal current flow direction and include dunes, sandwaves and ripples. Like aeolian sand dunes discussed on pp. 64–71 (Figure 2.21), these bedforms may possess a cross-bedded structure which indicates the direction of the current that formed them.

The two categories tend to characterise higher and lower tidal current velocities respectively, but at both high and low velocity extremes a plane sediment surface may arise, unornamented by any bedforms. Under the influence of a symmetrical tidal regime mobile sediment usually develops bedforms in accordance with the prevailing element of the ebb–flood tidal cycle, so that, for example, the lee side of a ripple will point shoreward on the flood and re-orientate to seaward on the ebb. However, under an asymmetrical tidal regime the velocity of a subordinate current may not be sufficient to completely reverse a bedform created by the dominant current. For example, a sand wave or ripple created by a dominant current would be draped by mud at slackwater; some sediment remobilisation may occur during the subordinate current phase, particularly reworking ripple crests along what is termed a reactivation surface, although its original orientation would be preserved. Deposition of this reworked sand occurs on top of the reactivation surface and previous mud drape where it still persists. A second mud drape may then be deposited at slackwater on top of the thin reworked sand layer. Tidal couplets preserved in this situation would be cross-bedded alternating thick and thin sand layers or bundles separated by mud drapes (Reading and Collinson 1996).

systems (tidal inlets and estuaries may silt up as a consequence) and net seaward transport in ebb-dominated systems.

Estuaries

Estuaries are perhaps the best known of tide-dominated coastal systems. They are often semi-enclosed with a restricted opening to the sea and commonly occur where the sea has invaded and drowned valleys and lowlands following the post-glacial rise in sea level. It is here that mixing between saline sea water and fresh river water occurs. Estuaries are host to a wealth of wildlife that exploit environments within the intertidal and subtidal zones, but estuaries are also important sites for human settlement and industry, and as such experience increasing environmental stress.

A large number of definitions for estuaries have been suggested over the past fifty years, but recently Dyer (1997) has incorporated earlier definitions: 'An estuary is a semi-enclosed coastal body of water which has free connection to the open sea, extending into the river as far as the limit of tidal influence, and within which sea water is measurably diluted with fresh water derived from land drainage' (p. 6). Although Dyer states that this is the most satisfactory overall definition, it is slightly ambiguous regarding river sections that are tidally influenced, but are beyond the upstream limit of sea water incursions. This definition acknowledges the importance of tides and salt–fresh water mixing in estuaries, but ignores sedimentological and biological aspects, which are included in definitions by Dalrymple *et al.* (1992) and Perillo (1995) respectively.

Estuary classification

A number of schemes have been proposed to classify estuaries, but most schemes are based on different aspects of the estuarine system, such as tidal range, topography, morphology, sedimentology, salt and fresh water mixing and circulation (Dyer 1997), and few attempt integrated classifications (e.g. Reinson 1992). Some schemes employ complex quantitative techniques which are beyond the scope of an introductory text like this. The following sections concentrate on outlining the schemes which are useful in the field and in understanding sediment dynamics within the estuarine system.

Estuary morphology

The shape of estuaries is very varied, but all possess:

- an estuary head, where the river enters the estuary;
- estuary margins that define the sides of the estuary; and
- an estuary mouth, where the estuary is open to the sea.

Following post-glacial sea level rise many valleys have been inundated and drowned by the sea. An estuary occupying a former river valley is known as a **ria**, and will retain the meanders, tributaries and other features associated with the original watercourse. The same is true for a former glaciated valley, which when drowned becomes a **fjord**. In both rias and fjords the rate of relative sea level rise outpaces the sediment infilling of the drowned valley, so that estuary morphology remains determined by the sides of the former valley and an **open-ended estuary** mouth persists. If sediment supply is high and the rate of deposition approaches being equal to the rate of sea level rise, then depositional features, such as salt marshes, occur around the estuary margins, narrowing the main tidal channel of the estuary. In many instances, where sediment is abundant within the estuarine system,

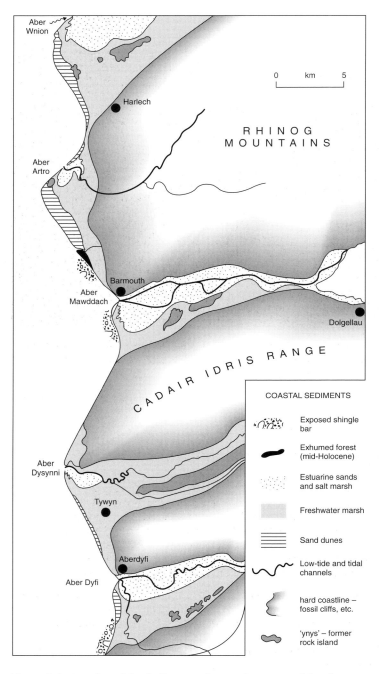

COASTAL SEDIMENTS

Exposed shingle bar

Exhumed forest (mid-Holocene)

Estuarine sands and salt marsh

Freshwater marsh

Sand dunes

Low-tide and tidal channels

hard coastline – fossil cliffs, etc.

'ynys' – former rock island

Figure 3.9 *A series of bar-built estuaries on the coast of Cardigan Bay, Wales, UK.*
Source: from Briggs *et al.* (1997) (Figure 17.13, p. 315)

a bar or spit may be built up by waves and longshore currents to partially obscure the estuary mouth to become a **bar-built** or **partially-closed estuary** (see also pp. 53–56) (Figure 3.9). Lagoons often occur just inside the mouth of a bar-built estuary, and occasionally the bar may extend to seal off the estuary mouth completely to form what is known as a **blind** or **lagoonal estuary**. Tidal current velocity is usually high in the constricted mouth of a bar-built estuary and coarse sediment brought into the estuary on these high velocity currents during the flood tide will be deposited just inside the mouth as a **flood-tide delta** (see Plate 3.2). In temperate regions, extensive flood-tide deltas may develop during the summer months, often causing navigation problems within the estuary. However, these sediment build-ups are usually flushed out of the estuary during times of heavy river discharge associated with winter storms.

Plate 3.2 *A downstream view of a road embankment built across L'Aber Estuary (Brittany, France). Note the flood-tide delta on the upstream side of the embankment, and the extensive tidal flats accreting on the downstream side.*

Box 3.3

Human interference in bar-built estuaries

Bar-built estuarine systems are particularly sensitive to human interference. An illustration of this is provided by ill-considered road building preferences in some regions. For example, in Brittany (northwest France) engineers have been inclined to construct embanked roads across the middle of bar-built estuaries, rather than bridges. Small fixed bridges are incorporated into embankments to allow river water to reach the sea, but these seldom constitute more than 3–4 per cent of embankment length. Apart from restricting navigation, this has a number of environmental consequences:

1   The embankment acts like a dam, generally increasing upstream water levels. In some estuaries, such as L'Aber Estuary on the Crozon peninsula (Plate 3.2), this has caused waterlogging of adjacent lowland soils leading to the demise of the woodland it supported and general habitat change.
2   Flood tides flowing through the bridge deposit deltas on the upstream side, which can further reduce river discharge through the bridge.
3   Sediment introduced and deposited into the outer estuary, downstream of the embankment, is protected and no longer flushed out of the estuary during periods of heavy river discharge. In this way the estuary becomes a sediment sink and continued

continued

sediment accretion leads to the progressive formation of tidal flats and salt marshes and ultimately terrestrialisation.

4 The bars themselves are also affected, because these bars usually comprise dunes that rely on sediment flushed out of the estuary to supply and maintain the integrity of the dune system. With sediment trapped in the estuary, dune fronts become eroded and vulnerable. This situation applies to the bar dunes of L'Aber Estuary, and more seriously at Lesconil in southern Finistère where the emerging tidal flats and salt marshes behind the bar have been reclaimed and developed, and are now at risk due to the retreat of the bar dunes that protect the area.

5 The residency time of pollutants within the estuary is increased exacerbating the problem of eutrophication.

Salt and fresh water mixing within estuaries

The manner in which salt water of sea origin mixes with fresh water of river origin has implications for understanding estuarine sedimentology and provides a convenient means of classifying estuaries regardless of morphology. However, classifying estuaries according to mixing requires detailed measurements of the salinity structure of the estuarine water column, and is therefore not so readily utilised in the field without specialised equipment. Pritchard's (1955) seminal work on estuarine circulation has been the foundation for much subsequent research into estuarine hydrodynamics and forms the basis of the classification described below.

Pritchard considered estuaries using a **salt-balance principle** which states mathematically that the rate of salinity change at a fixed point within an estuary is brought about by the operation of two processes:

1 **diffusion** occurs when the differences in the ionic composition of salt and fresh water produces turbulence that mixes the water in the estuary to ultimately attain a uniform salinity; and

2 **advection** involves the physical mixing between salt and fresh water due to internal circulation.

Although these processes co-occur, one of them may dominate mixing in a given estuary. It is worth emphasising the difference, in that diffusion may be regarded as chemically-driven mixing, whilst advection represents physical mixing. The manner and degree of mixing in an estuary determine its type according to Pritchard's classification (Figure 3.10).

Stratified estuary

In estuaries where the mixing of salt and fresh water is minimal, the water column becomes stratified, with a lower high salinity layer and an upper fresh layer (Figure 3.10a). The layer sequence is determined by density differences, with denser sea

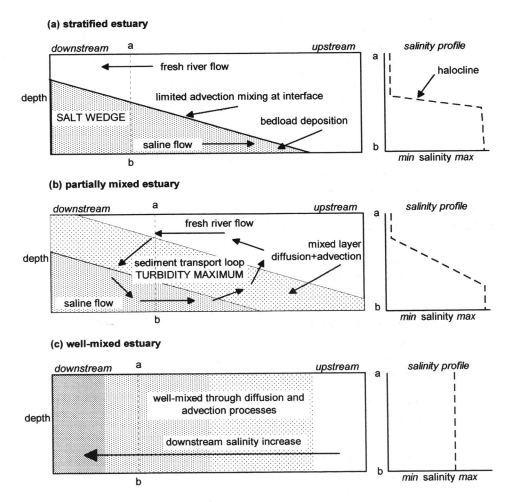

Figure 3.10 *Estuary types according to Pritchard's (1955) salt-balance principle classification: (a) stratified estuary; (b) partially mixed estuary; and (c) well-mixed estuary.*

water occurring below the lighter and buoyant fresh water. Stratified estuaries are most common along protected microtidal coasts, and because of this the relative movement of the two flows, even during the ebb tide phase, is almost always in opposite directions, so that the lower saline layer deforms to offer least resistance to the faster overflowing river water. In doing so, the saline layer thins or tapers upstream to form a **salt-wedge**. The interface between the two flows is known as the **halocline**, because it is here that salinity changes abruptly. As with all interacting fluids of differing densities (Figure 2.7), waves form at the interface and these waves allow some limited physical mixing (i.e. advection) to occur along the halocline. Sediment behaviour within such an estuary is strongly influenced by the hydrodynamic conditions. Suspended sediment of river origin is often retained within the upper fresh layer until the open sea is reached, where a sediment plume extending out to sea may be observed. Sediment transported as bedload, however, whether of river or marine origin, is most often deposited at the tip of the salt-wedge, as it is here that flow velocity is either reduced or direction reversed depending on the state of the tide. The position of this bedload deposition shifts up and downstream in response to changes in tides and river discharge, the latter being the most important. During episodes of high river discharge, such as during floods, these bedload deposits may be flushed out of the estuary entirely. Fjords are also examples of stratified estuaries, but lack salt-wedges. Stratified estuaries have been variously referred to as Type A, river-dominated, salt-wedge (or fjord) estuaries. A well-known example of a stratified estuary is that of the Mississippi River in the USA.

Partially mixed estuary

These are more influenced by tides than stratified estuaries and are typical of mesotidal to macrotidal settings. Tidal turbulence, caused by the ebb and flood entering and exiting the estuary, destroys the interface between the salt-wedge and overlying fresh water to produce a more gradual salinity gradient through the water column (Figure 3.10b). Both advection and diffusion processes operate, and large density-driven eddies exist which help to exchange salt water upwards and fresh water downwards. A broad mixed salinity zone is created, corresponding to the steepest part of the vertical salinity gradient (i.e. halocline), which shallows downstream, so that it is at the bed near the estuary head and at the water surface near the estuary mouth. This inclined mixed zone, therefore, is slightly reminiscent of the salt-wedge morphology, occurring between the opposing river and tidal flows. Thus, a two-layer flow exists, separated by a **level of no motion** where the average flow velocity is zero, and which approximately corresponds to both the position of the mixed zone and the halocline. Also, the location on the estuary bed where the tidal and river flows meet and converge is known as the **null point** (Dyer 1997).

The sediment dynamics of a partially mixed estuary are substantially different to those in a stratified type. Suspended sediment transported downstream in the river flow will at some point encounter the mixed zone and its associated lower flow velocity. Particles will begin to settle out, but once below the mixed zone they become entrained and transported back upstream by the tidal flow. The same particles may then re-encounter the mixed zone and be mixed upwards by eddies into the river flow. Particles may circulate in a closed-loop like this for some time, and the high concentration of suspended sediment trapped in this way around the mixed zone is known as a **turbidity maximum**. The position of the turbidity maximum moves up and down the estuary with the ebb and flood of the tide. Partially mixed estuaries have also been referred to as Type B estuaries, and a good example is the Mersey in northwest England.

Well-mixed estuary

This estuary type is dominated by tidal activity and requires severely macrotidal and hypertidal conditions to effectively mix the waters, through both advection and diffusion processes. A well-mixed estuary, or vertically homogeneous estuary, characteristically lacks a vertical salinity gradient, so that salinity is uniform from surface to bed at any given point within the estuary (Figure 3.10c). However, there are three subdivisions within this category:

1 Where the estuary is particularly wide the Coriolis effect may separate the flows so that the seaward river flow is restricted to the right side of the estuary (in the northern hemisphere) and the landward tidal flow to the left, but within each flow salinity remains uniform throughout the water column. This type has been referred to as **laterally inhomogeneous estuary**.

2 Some estuaries have no separation of river and tidal flows, and in such cases sufficient mixing may give rise to uniform salinity at all points across an estuary, so producing a **laterally homogeneous estuary** (also known as a sectionally homogeneous or Type C estuary by some authors). However, salinity does change with distance down the estuary, so that minimum salinity occurs at the head and maximum salinity towards the mouth of the estuary.

3 Where salinity is uniform both laterally across an estuary and longitudinally along the length of an estuary, from head to mouth, then a truly **homogeneous estuary** (or Type D estuary) is defined. This type constitutes the theoretical end member of the spectrum of estuary types and opposes the highly stratified estuary described above.

Estuarine sedimentation

The sediment within an estuary comes from one of three main sources:

1 **Fluvial/glacial sources.** This is sediment supplied by the river(s) flowing into an estuary, and/or from glaciers along cold coasts. It is considered to be of terrestrial origin and has generally increased through historic times with the advent of widespread farming, which has stimulated soil erosion in many river catchments. The supply of this sediment may be seasonal, often being associated with high precipitation events during winter storms.

2 **Estuary margin sources.** This describes sediment eroded from the margins of the estuary itself. The material being eroded may either be soft sediment previously deposited by the estuary, or a geologically older material, such as the local bedrock. Tidal currents working within the estuary may be sufficient to rework soft sediment, but erosion increases with occasional high wave energy activity within the estuary. Indeed, significant erosion of hard bedrock requires these increased energy levels.

3 **Extra-estuary sources.** This encompasses sediment supplied from outside the estuary mouth, and includes sediment eroded from cliffs along the coast downdrift of the estuary mouth and continental shelf sediment thrown into suspension by passing waves. Near the estuary mouth this material is entrained by flood tidal currents and transported into the estuary.

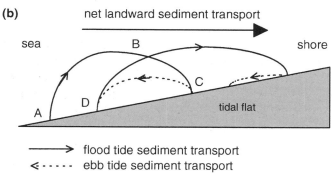

Figure 3.11 *(a) Sediment pathways in an estuarine system; and (b) the mechanics of settling lag (see text for explanation) and its role in inducing net landward sediment transport.*

Once the sediment is supplied to the estuary from one of the various sources, its fate is determined by the combined effects of sediment particle size, wave–tide–river energy, and the salt–fresh water mixing relationship within the estuary (Figure 3.11a). Sediment composition in an estuary that is dominated by fairly coarse sand-grade material is known as a **non-turbid estuary**, so named because there is a low concentration of **suspended particulate matter** (SPM) within the water column. The sand may be transported in suspension during periods of maximum current velocity, but quickly settles out following the peak tidal flow, leaving the water relatively clear (i.e. non-turbid). A **turbid estuary**, however, is characterised by fine sediment composition of mud- to silt-grade. These particles are easily held in suspension and so the concentration of suspended particulate matter is usually high at all states of the tide. The deposition of these fine particles is aided by the process of **flocculation**, which occurs when clay particles in the river water come into contact with salt nucleii upon mixing in the estuary. This process encourages clay particles to join together or coagulate to form larger particles called **flocs**, which are heavier and so more readily deposited. Fine sediment may also be ingested by invertebrates and exported to the estuary bed as faecal pellets. Carter (1988) describes the behaviour of suspended sediment in a turbid estuary through a tidal cycle. He suggests that at peak tidal flow sediment throughout the water column is well-mixed, producing a uniform SPM concentration at all depths. As the tidal flow decelerates SPM starts to become more concentrated at depth, producing a sediment concentration gradient with depth, known as the **lutocline**. With progressive deceleration of the tidal flow the lutocline may become stepped or stratified, until finally these layers merge at or above the bed at the time of slackwater. The sequence is then reversed as the tidal flow accelerates following slackwater.

Sedimentary environments and particle size gradients within an estuary have been described by Dalrymple *et al.* (1992). They divide estuaries longitudinally, according to energy sources and their relative amounts, into a marine-dominated outer section near the mouth, a mixed energy central section around the estuary mid-point, and a river-dominated inner section near the head. The outer and inner sections have high energy levels related to marine and riverine input respectively, whilst the central mixed energy section generally possesses the lowest energy. This is because marine and riverine energy decreases up and down the estuary respectively, both being relatively low in the central section. The sedimentological consequence of this energy partitioning is a saddle-shaped particle size gradient along the estuary, with relatively coarse sediment occurring at both the high energy head and mouth regions, and finer sediment in the lower energy central section. Dalrymple *et al.* also suggest an estuary classification based on the dominant marine process, so that they recognise distinct **wave-dominated** and **tide-dominated estuaries**. Reinson (1992) states that wave-dominated estuaries can occur in microtidal to low macrotidal settings of stratified to partially-mixed type, and morphologically include blind and bar-built to open-ended estuaries. Tide-dominated estuaries, however, only occur under high macrotidal to hypertidal conditions, associated with well-mixed and open-ended estuary types.

In addition to particle size variation along an estuary, size variation also occurs laterally across an estuary. Energy tends to decrease from the central estuary axis to the intertidal margins, and this corresponds to a similar decrease in particle size. Fine sediment tends to be transported landward and deposited close to the margins of the estuary due to the phenomena known as **settling lag** (Postma 1961), which contributes to the formation of depositional features such as mudflats, salt marshes and mangroves. Settling lag works when a flood tidal current attains a threshold velocity sufficient to entrain a given sediment particle at location A, which is subsequently transported landward in the flow. The velocity of a flood tide current decreases landward, in addition to the approach of slackwater, so that the velocity eventually falls to the entrained particle's **settling velocity** threshold at B. The particle does not fall vertically to the sediment surface at B, but is carried by inertia further landward to C. The particle is re-entrained by the ebb at C, but at a much later stage in the ebb cycle. Therefore, the particle is transported seaward for a shorter time period and is deposited at D, which is landward of its original position A (Figure 3.11b).

Tidal sand and mudflats may develop then as a direct consequence of settling lag, eventually building up to allow vegetation to take hold to become salt marsh or mangrove (see pp. 97–108). However, in arid environments a particular type of tidal flat environment may develop called **sabka**. These are a distinct coastal environment that characterise arid regions where they replace the more familiar mudflats, salt marsh and mangroves (Box 3.4).

Box 3.4

Sabkas – tidal flats of arid environments

These are low gradient salt flats, with extensive examples found in the Arabian Gulf, Egypt, and Mexico, and they occur both in the intertidal zone and supratidally. Mangroves are sometimes associated with sabkas, but most frequently they support algal mat communities that are tolerant of the arid and saline conditions. Salts often precipitate to form hard salt crusts on the surface, and other evaporite minerals are also common, such as gypsum (Viles and Spencer 1995). Both clastic and carbonate sabkas exist, the former being associated with deltaic sedimentation, such as at Bahrah in northern Kuwait, part of the Tigris–Euphrates Delta (Saleh *et al.* 1999). The Bahrah sabka is approximately 5.5 km wide and is apparently unique in that it comprises a landward sediment zone characterised by terrigenous quartz silt and sand deposited through aeolian processes, and the more usual seaward carbonate and evaporite deposits. The development of the extensive Bahrah sabka has taken place since 3,040 years ago, when the sea levels of the region stabilised, growing seaward at a rate of 1.5 to 2 m per year.

Estuarine ecology

With the exception of high intertidal salt marshes and mangroves, which will be discussed later, estuaries typically lack abundant macroflora. This is because the subtidal and low intertidal sediment substrate is highly mobile, and only occasional stones or gravel patches will support algal colonies. Higher up in the intertidal zone the environment is less energetic and as a consequence the sediment is less mobile. Under such conditions, oxygen is depleted rapidly within the sediment profile leading to anoxic conditions that support hydrogen sulphide-producing bacteria. This results in pungent, thick black mud being created just under the active sediment surface. A rich community of microflora and microfauna, such as diatoms and foraminifera, exists in this environment, as do a number of burrowing higher order organisms, such as ragworms and various molluscs. This infauna in turn supports diverse and abundant populations of fish and wading birds.

A significant environmental parameter to affect estuaries, more so than most other coastal environments, is salinity. Salinity can range from fresh water to normal salinity sea water of approximately 33 per cent salt content. Although any given point in an estuary experiences a range of salinities throughout the tidal cycles, the longitudinal salinity gradient down an estuary produces a marked ecological zonation, particularly apparent in invertebrates (Cremona 1988).

Salt marshes and mangroves

For most of the tidal cycle, the upper part of the intertidal zone is exposed to subaerial conditions, and it is here in mainly tide-dominated situations that salt-tolerant plants may grow to create widespread vegetated intertidal surfaces, known as either salt marshes or mangroves. They also develop on deltas, but these are mainly river-dominated and will be discussed in the next chapter. Salt marshes are characterised by short plants, such as grasses, and are mostly restricted to temperate coastlines, whilst mangroves are their tropical and subtropical counterparts, and comprise trees of various heights which can develop into extensive mangrove forests or mangals. The vegetation is often zoned from a zone of pioneer species seaward grading landward to a more mature community. In salt marshes, these vegetation changes are used to subdivide the environment into low and high marsh respectively, although these again may be separated by an additional intermediate or middle marsh zone. Mangroves in tide-dominated settings may also display a vegetation zonation. Both have common aspects to their geomorphology and sedimentology, but distinct differences make separate treatment here logical.

Salt marsh geomorphology

The large-scale geomorphological impression of these coastal systems is that of a near- or quasi-horizontal platform, that slopes gently seaward. However, on smaller scales a wide variety of features may be seen, such as cliffs, salt pans, tidal creeks and their levées. The platforms are built up by the deposition of sediment being brought on to the marsh surface by flood tide currents and trapped by vegetation (although, see Viles and Spencer (1995: 168) who reiterate that vegetation may not be as important, on some salt marshes for trapping sediment, as was previously considered). In this way, sediment **accretion** on the surface of a marsh leads to its elevation within the tidal frame; it is worth noting that the rate of surface elevation is usually less than accretion, because accreted sediment often compacts under the weight of successive episodes of deposition. As will be seen in the next section, the rate of accretion varies across a marsh transect, so that relatively more sediment is accreted lower in the tidal frame than near the high tide limit. This process ensures that lower surfaces elevate at a faster rate, so as to catch up the initially higher surfaces, resulting in a quasi-horizontal platform. Salt marshes may exist on a wide variety of tidally influenced coastlines. French (1997) illustrates seven topographically distinct salt marsh settings, including those found along open coasts and embayments, those protected behind barriers, and those that occur along the fringes of estuaries (Figure 3.12).

Features marking the boundary between marsh platforms and seaward mudflats are variable. Allen (1993) documents three major categories of marsh–mudflat boundaries that he recognises along turbid British tidal coasts, although the categories can be widely applied (Figure 3.13).

1 **Ramped marsh shore.** This is a gently sloping, smooth transition from mudflat to marsh, and may exist where sediment supply is abundant. In these conditons, the marsh is actively accreting seaward and indicates a positive shore regime.
2 **Cliffed marsh shore.** Under normal circumstances very little vertical erosion of sediment occurs from the marsh/mangrove surface, but this does not include sediment resuspended from the layer at the base of the lutocline at slackwater, which fails to become deposited. Quite often, however, cliffs up to several metres high occur at the edge of salt marshes, commonly separating the marshes from seaward mudflats. These cliffs are evidence that erosion is taking place in the horizontal plane and that the marsh is retreating landward, indicating a negative shore regime. Allen (1989) documents the principal failure mechanisms for marsh cliffs, which includes toppling failure and rotational slumping (see Figure 2.10). Erosion and cliff development are associated with the meandering of tidal channels, increases in wave activity, and decreases in sediment supply, although once initiated wetting–drying cycles and seasonal changes in climate help to perpetuate cliff retreat (Plate 3.3).
3 **Spur and furrow marsh shore.** This is the least common of the shore types, although may be locally widespread. It is characterised by a generally ramped shoreline that is dissected by numerous furrows, which are separated by ridges or spurs. The furrows are products of very localised erosion, which upon inititation trap pebbles, gravel and sand which get washed up and down the furrow by the tide, thus aiding further erosion and furrow deepening. Deposition occurs on top of the spurs, and in this way furrow erosion and spur-top deposition is balanced, with neither retreat nor advance dominating, indicating a neutral shore regime.

Tidal creeks dissect salt marshes and are the principal conduits through which tidal water and sediment floods on to and ebbs off the salt marsh surface. Morphologically, tidal creeks are similar to river channels in that they often develop dendritic networks and meandering channels complete with levées. However, the hydraulic and flow conditions differ from rivers, because whereas rivers are at their most energetic when the channel is full (i.e. under storm conditions), tidal creeks are at their most sluggish at bank full, because this usually occurs at slackwater associated with high tides, when flood flows reverse to become ebb flows. Under fairweather conditions, the ebb tide appears to be dominant in creek channel and headwater erosion. Commonly a cascade of smaller channels may be seen

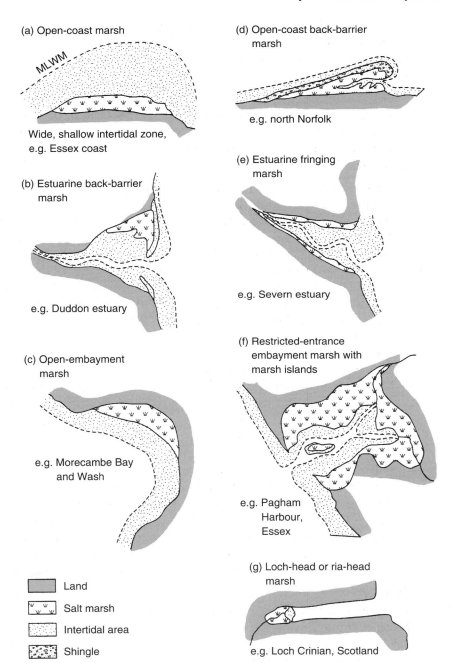

(a) Open-coast marsh

MLWM

Wide, shallow intertidal zone,
e.g. Essex coast

(b) Estuarine back-barrier
marsh

e.g. Duddon estuary

(c) Open-embayment
marsh

e.g. Morecambe Bay
and Wash

(d) Open-coast back-barrier
marsh

e.g. north Norfolk

(e) Estuarine fringing
marsh

e.g. Severn estuary

(f) Restricted-entrance
embayment marsh with
marsh islands

e.g. Pagham
Harbour,
Essex

(g) Loch-head or ria-head
marsh

e.g. Loch Crinian, Scotland

Land

Salt marsh

Intertidal area

Shingle

Figure 3.12 *Various salt marsh settings in tidal environments.*
Source: from French (1997) (Figure 2.11, p. 46)

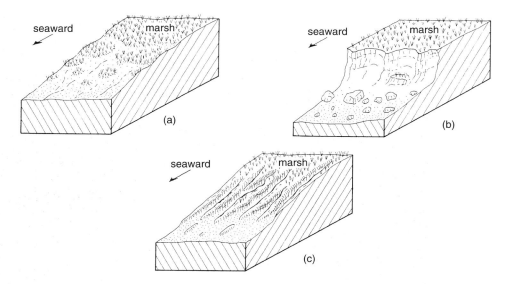

Figure 3.13 *Three categories of salt marsh shorelines: (a) ramped; (b) cliffed; and (c) spur and furrow.*

Source: from Allen (1993) (published with the permission of the Geologists' Association)

Plate 3.3 *Extensive cliffing of a salt marsh at Northwick Oaze in the Severn Estuary (UK). The sediment corer is c. 1.25 m long.*

Plate 3.4 *A nest of channels in a salt marsh tidal creek at Barneville (Normandy, France). The salt marsh vegetation is mainly sea purselane (*Halimione portulacoides*).*

embedded in a creek cross-section, each representing erosion during later stages in the ebb cycle under decreasing discharge (Plate 3.4). However, during storm conditions, the flood tide may become dominant, with high flow velocity and discharge directed landwards within the creeks (Bayliss-Smith *et al.* 1979; Knighton *et al.* 1992).

Salt pans are unvegetated pools of standing water that occur on the salt marsh surface. There are two main morphological types, roughly circular 'primary' pans, and elongate channel pans. Some controversy surrounds the origin of 'primary' pans, as some authors consider them to develop from the outset of salt marsh formation (hence 'primary' pans) as an expression of uneven accretion on the surface, so that areas of low accretion develop into pans. Others have suggested that they are formed by tidal litter scouring hollows which develop into pans. The origin of channel pans, however, appears more secure in that they represent tidal creeks that have been severed by fallen sediment blocks creating a dam which ponds up ebb tidal water. Whatever their origin, salt pans are conspicuous features of most salt marsh systems, and play an important ecological role.

Salt marsh sedimentology

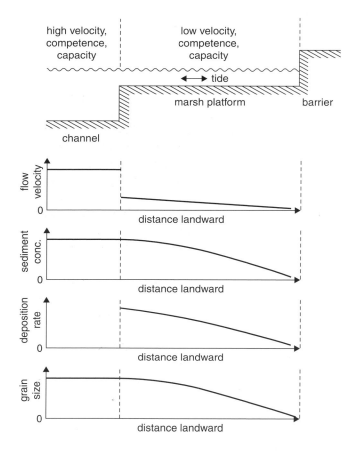

A number of models have been proposed to describe the sedimentation on salt marshes. All are relatively similar, often employing daunting expressions. Allen (1994) proposes such a model, but thankfully later summarised it in a more elementary form (Allen 1996). Although the details of the model are not appropriate here, the results of the model are useful as aids to understanding general principles of salt marsh sedimentation and subsequent development (Figure 3.14).

Figure 3.14 *A model predicting sediment deposition and distribution upon quasi-horizontal salt marsh surfaces.*
Source: from Allen (1996) (published with the permission of the Geologists' Association)

- The model simplifies the real world by assuming a constant morphology for the salt marsh shore, in that a horizontal salt marsh platform is bounded to the landward side by a barrier, such as an artificial sea wall or natural cliff and to the seaward side by a tidal channel, which could be a tidal river in an estuarine setting. The tidal channel is the source of both water and sediment delivered on to the marsh platform.

- The model predicts that at any time when the salt marsh surface is submerged, the velocity of the tidal flow decreases linearly landward, whether associated with the flood or ebb tides. The model takes no account of friction or turbulence, so that the contrast between the high velocity channel and low velocity platform is solely due to differences in the momentum, competence and capacity of the two environments.

- The landward decrease in flow velocity has three major sedimentological consequences: (1) reduced flow velocity allows relatively coarse sediment to be deposited shortly after entering on to the marsh platform, so that the model predicts a landward decrease in sediment grain size across the marsh; (2) this rapid deposition of coarse sediment near the marsh edge contributes to a predicted overall landward decrease in deposition rate; and (3) the sediment concentration of the tidal water across the marsh is also predicted to decrease landward, a consequence which is closely related to the above two points.

Salt marsh morphodynamics

Like most coastal sedimentary systems, salt marshes are extremely sensitive to changing environmental conditions. Changes in sediment accretion and erosive processes are amongst the principal controls on salt marsh morphodynamics, and in particular in determining the position of salt marsh shorelines. Figure 3.15 shows the relationship between these variables and indicates that during periods where the influence of erosion is greater/lesser than the horizontal accretion of sediment, the marsh shoreline will retreat/advance (negative/positive shore regimes). Where the regime alternates between negative and positive, then successive phases of erosion and accretion may occur. These phases manifest themselves

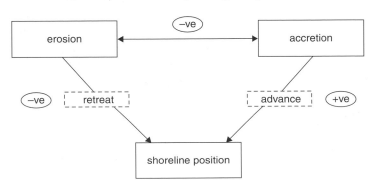

Figure 3.15 *The relationship between horizontal erosion and accretion in determining salt marsh shore regime; where accretion > erosion then the shore will advance (positive shore regime), and where accretion < erosion the shore will retreat (negative shore regime).*

as a series of seaward descending terraces, each separated by small clifflets, which are the protruding remnants of cliffs developed in the retreating phases, and then partially buried by subsequent accretion. Allen (1993) describes these as offlapping morphostratigraphic units, and notes their widespread occurrence in estuarine fringing marshes (Figure 3.16). In the hypertidal Severn Estuary in southwest Britain, at least three major morphostratigraphic salt marsh units have been recognised and given full geological formation status. The oldest is the Rumney Formation which began accreting in the sixteenth century, then the eighteenth-century Awre Formation, and finally the Northwick Formation which started forming in the twentieth century. A period of horizontal erosion and shoreline retreat stratigraphically separates each formation, but each formation continues to vertically accrete sediment from the outset.

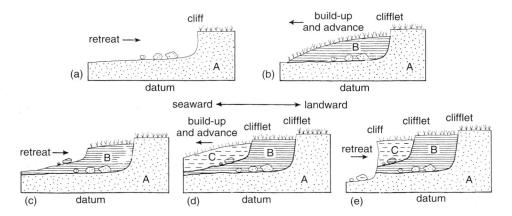

Figure 3.16 *Alternating negative and positive salt marsh shore regimes result in the formation of seaward descending offlapping morphostratigraphic units.*
Source: from Allen (1993) (published with the permission of the Geologists' Association)

Salt marsh ecology

Salt marsh geomorphology is closely linked to ecology, because it is widely accepted that salt marsh plants make a significant contribution to the trapping and accretion of sediment. They also produce large amounts of particulate and dissolved organic matter which play a part in coastal food web dynamics. Salt marshes first develop when unvegetated mud or sand flats become colonised by pioneering **halophytes**, which include plants such as *Spartina* (cord grass) and *Salicornia* (glasswort) species. This low salt marsh community has to withstand extreme environmental pressures, including the instability of the sediment and its high salt content, regular tidal submergence with its associated waterlogging effects and minimal oxygen availability, as only the sediment surface is aerated. The upper salt marsh sediment is more stable, less frequently inundated by the tide and so is able to support a more diverse plant community. In Europe, typical upper salt marsh plants include *Aster tripolium* (sea-aster), *Plantago maritima* (sea-plantain), *Limonium vulgare* (sea-lavender), *Cochlearia officinalis* (scurvey grass), *Halimione portulacoides* (sea-purselane) and *Puccinellia maritima* (salt marsh grass). Also, close to the landward limit of tidal inundation, the strandline is characterised by plants less tolerant of salt, such as *Juncus maritimus* (sea-rush) and *Phragmites communis* (common reed). Salt marshes then, possess a clear plant zonation reflecting vegetation colonisation and succession which is strongly influenced by factors associated with the frequency and duration of tidal inundation. Similar zonations are also exhibited by other organisms, such as diatoms and foraminifera. Animals that are found on salt marshes are typically terrestrial species at the limit of their range. However, the more marine lower salt marsh may be home to some marine animals, such as estuarine crabs (e.g. *Carcinus maenas*) and snails

Box 3.5

Reconstructing salt marsh shoreline position

Using Allen's sedimentology model described above, Allen (1996) is able to reconstruct historical variations in the shoreline position of salt marshes in the Severn Estuary. This is possible because at any fixed location on a salt marsh particle size will increase through time as the shoreline retreats landward, and decrease as the shoreline advances seaward (Figure 3.17). Therefore, according to the model, particle size variations in a salt marsh sediment sequence reflect changes in the position of the shoreline. This model is, therefore, potentially very useful in investigating the extent and rates of salt marsh morphodynamic changes.

Figure 3.17 *The influence of salt marsh shoreline position on temporal grain size distribution.*

Source: from Allen (1996) (published with the permission of the Geologists' Association) (e.g. *Hydrobia* species). Furthermore, because salt marshes are productive throughout the year, they are often important wintering sites for wildfowl (Cremona 1988).

Mangroves

The principal difference between salt marshes and mangroves is the greater above ground biomass in mangroves. There are many different species of mangrove trees, but amongst the commonest genera seen at a mangrove shore are *Rhizophora*, which possess dense networks of prop roots that extend into the sediment surface from the above ground trunk and *Avicennia* that have a horizontal below ground root system that sends up shoots called pneumatophores (Plate 3.5). Both strategies assist in anchoring the trees to the substrate, and also the partial above ground roots help to facilitate oxygen intake, as the substrate is invariably anaerobic. Such root networks are also considered to promote the trapping and deposition of sediment that is introduced into the mangal system, up to approximately 2 mm y^{-1}.

Salt marshes and mangroves are geographically distinct, with salt marshes dominating in the temperate zones, whilst mangroves are restricted to lower latitudes between 32°N (Bermuda) and 38°S (Australia). Zonation of mangrove species does occur with *Avicennia* and *Rhizophora* often characterising the seaward zones, whilst further inland species of other genera such as *Laguncularia* appear. Three main physical mangrove settings occur (Woodroffe 1993):

1 **River-dominated setting** – this setting refers to substantial mangroves that exist on many low latitude deltas, where the sediment is principally supplied by river water, for example, the Fly River Delta in Papua New Guinea.
2 **Tide-dominated setting** – mainly associated with estuarine situations in which sediment is supplied by tides and tidal currents, for example, the South Alligator River in the Northern Territory of Australia.
3 **Carbonate setting** – restricted to coral reef settings where mangroves may grow around cays and in protected areas behind storm shingle ramparts and coral islands. The sediment here is of local derivation being supplied by wave erosion of coral. However, if this supply is limited the mangroves may develop organic-rich peat deposits, for example, Grand Cayman.

Mangroves are ecologically rich and are refuges for many species of birds and other animals. Also, the sheltered mangrove environment, particularly creek networks, are known to act as nurseries for a number of fish and sea food (shrimp and prawn) species, some of which are important food and profitable species for local communities. Furthermore, dense mangals offer significant protection to coastal communities from tropical storm waves and associated erosion. However, this fact is often neglected and many mangrove sites in developing countries are being converted to agriculture, such as rice cultivation, which is ironically less profitable than harvesting shrimp in undamaged mangroves. For example, in some west

Plate 3.5 *Mangroves of the eastern Australian coastline: (a) Avicennia mangroves with abundant pneumatophores (Minnamurra Estuary, New South Wales); (b) Rhizophora with prop root networks (Thomatis Creek, Queensland).*

African states original mangroves have declined in area by 46 per cent, and in the Philippines by as much as 75 per cent (French 1997).

Summary

- Tides are generated by the combined gravitational attraction of both the moon and the sun upon the earth, with different orbital configurations giving rise to spring and neap tides. Predicted tidal levels can be significantly altered by meteorological effects.

- Tidal range strongly influences the type of landforms present in a tidal coastal system, as it determines the spatial extent of marine activity at the coast and influences the type and operation of weathering, erosional and depositional processes.

- Tidal environments, such as estuaries, salt marshes, mangroves and sabkas, possess distinctive ecosystems that are often under pressure from human society. For example, many estuaries are centres for urban settlement and industry.

Discussion questions

1 Evaluate the geomorphological significance of tidal range.

2 Assess the suitability of salt–fresh water mixing for the classification of estuarine systems.

3 Salt marshes are dynamic tidal coastal systems. Examine the dynamic relationship between the geomorphology and sedimentology of a temperate salt marsh.

Further reading

See also

Tide-dominated deltas, Chapter 4.
Coastal responses to sea level change, Chapter 5.
Coastal management issues, Chapter 6.

General further reading

Tides, Surges and Mean sea level. D. T. Pugh. 1987. Wiley, Chichester, 472pp. A fairly detailed account of tides, tidal predictions, levels and meteorological effects.

Estuaries: A Physical Introduction (2nd Edition). K. R. Dyer. 1997. Wiley, Chichester, 210pp. An introductory text to estuaries, including their classification, tidal effects and mixing processes, but quite mathematical in places.

Estuarine shores: Evolution, Environments and Human Alterations. K. F. Nordstrom and C. T. Roman (eds). 1996. Wiley, New York, 510pp. An advanced compilation of work on the very varied nature of estuarine environments.

The Estuarine Ecosystem (2nd Edition). D. S. McLusky. 1989. Blackie, Glasgow, 215pp. A now standard and well-accepted introduction to estuarine ecosystems.

Saltmarshes: Morphodynamics, Conservation and Engineering Significance. J. R. L. Allen and K. Pye (eds). 1992. Cambridge University Press, Cambridge. A useful and wide-ranging introduction to salt marsh systems, their ecology and applied issues.

Coastal and Estuarine Management. P. W. French. 1997. Routledge, London, 251pp. Although aimed at being a general coastal management text, because of the author's interests it leans towards the management of tidal environments, especially estuaries.

Intertidal Ecology. D. Raffaelli and S. Hawkins. 1996. Chapman and Hall, London, 356pp. A timely general introduction to the ecology of intertidal environments along all types of coasts.

4 River-dominated coastal systems

River-dominated coastal systems are known as deltas, which are coastal accumulations of river-supplied sediment, although they also have complex interactions with waves and tides. They are often very large and act as centres for human settlement, agriculture and industry, but are also very sensitive to human interference. This chapter covers:

- **the classification of deltas and the role of rivers, waves and tides**
- **the hydrodynamic relationship between rivers and the sea within the delta system**
- **the sedimentological and geomorphological characteristics of deltas**
- **the human habitatation of delta systems and their sensitivity to interference**

Introduction

In recent years, delta systems have not been discussed in detail by coastal scientists. This is because deltas comprise many of the different coastal environment types that are associated with wave-dominated and tide-dominated systems, and because their distinctiveness comes from the input of water and sediment from riverine sources. However, they are major physical constructions that occur at the coast and are extremely important as locations for human settlement, agriculture and industry. Furthermore, they are dynamic systems, not only responding to changes in the river systems that supply water and sediment, but also to coastal processes: wave, tide and aeolian processes. They are fascinating coastal environments, often of regional if not global significance, and worthy of inclusion here. The balance is being addressed, however, within the academic coastal community, a move which was reflected in a special issue of the *Journal of Coastal Research* in 1998 devoted to deltas as coastal environments.

The physical nature of deltas has been mainly established, not by coastal geomorphologists but by sedimentologists working in the petroleum exploration industry. This is because deltaic environments, mostly ancient, contain many of the sedimentological criteria necessary for hydrocarbon (gas and oil) development. This includes organic deposits, such as peat and coal that may give rise to hydrocarbons (source rocks), and also vast sand bodies with abundant and sizeable pore spaces between the sand grains, playing host to hydrocarbons (reservoir rocks).

The ecology of deltaic systems is not covered in this chapter. This is because deltaic ecosystems are extremely varied from riverine habitats (which are not coastal), to estuaries, mangroves, salt marshes, sabkas, barrier islands, coastal sand dunes and beaches, all of which are covered elsewhere in this book. Therefore, this chapter concentrates upon the hydrodynamics, geomorphology, sedimentology and human influence in the delta system.

Delta classification

Traditionally, deltas have been categorised according to their overall plan morphology (Table 4.1). Since Galloway (1975), however, deltas have been classified according to the dominant process operating in an individual delta system (Figure 4.1), so that a delta may be described as river-dominated, tide-dominated or wave-dominated. These process relationships are illustrated by a ternary model (Figure 4.2), a triangular diagram with the three processes each occupying a corner of the triangle. Individual deltas are then plotted within the diagram according to the relative importance of the different processes.

River-dominated deltas

These extend outwards from the coast due to sediment-laden river water jetting out into the sea. This tends to form levées that define the river channels, and hence the

Table 4.1 *Traditional categorisation of deltas according to their overall plan morphology.*

Delta plan	Description	Example
Arcuate (fan-shaped) delta	Triangular-shaped delta, with downstream branching distributary channels to the sea	The Nile Delta (Egypt)
Bird's foot delta	Sediment builds up as levées either side of a channel, but deep water exists between channels	The Mississippi Delta (USA)
Compound delta	Deltas merging from more than one river outflow	The Ganges–Brahmaputra Delta (Bay of Bengal)
Estuarine delta	Drowned river valley infilled by the deposition of river sediment	The Mobile River Delta (USA)
Cuspate delta	A delta whose shoreline is extensively modified by wave action	The Rhône Delta (France)

Source: after Park (1997)

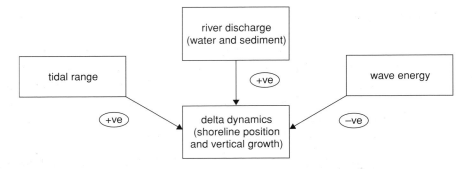

Figure 4.1 *The relationship between delta dynamics and river, wave and tidal influences.*

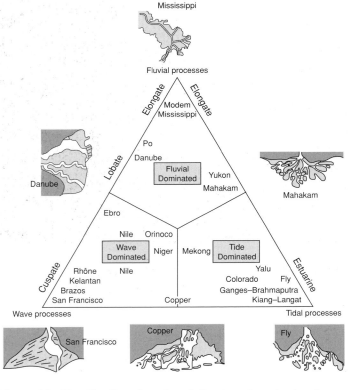

Figure 4.2 *Classification of modern deltas based on dominant processes (waves, tides and rivers).*
Source: from Briggs *et al.* (1997) (Figure 17.9, p. 311)

name **bird's foot delta** has been applied to this rather spindly morphology. The Mississippi Delta in the southeast USA is a good example.

The significance of riverine activity in delta dynamics has been highlighted by Hensel *et al.* (1999) in a study of the Rhône Delta along the Mediterranean coast of France. Sediment accretion and the resulting elevation of the delta surface have been measured between 1992 and 1996. Accretion in areas of the delta subjected to riverine flooding averaged 13.4 ± 7.0 mm y^{-1}, whereas areas open to marine conditions accreted only 1.2 ± 0.5 mm y^{-1}. The resulting surface elevation in the riverine areas was 11.3 ± 6.1 mm y^{-1}, with the difference accounted for by compaction of the accreted sediment. The accretion rate in the marine areas of the delta is less than the current rate of sea level rise for the region, so that these marine areas, which are so valuable to wildlife, are being submerged.

Tide-dominated deltas

These include deltas where sediment delivered to the coast by a river is redistributed by tidal currents. They possess tidal channels as well as river channels which creates a rather fragmentary finger-like shoreline. This can be further enhanced by parallel-orientated bedforms, such as tidal shoals, bars and inlets, situated within larger tidal channels (Plate 4.1). The Ganges–Brahmaputra Delta of India and Bangladesh falls within this category.

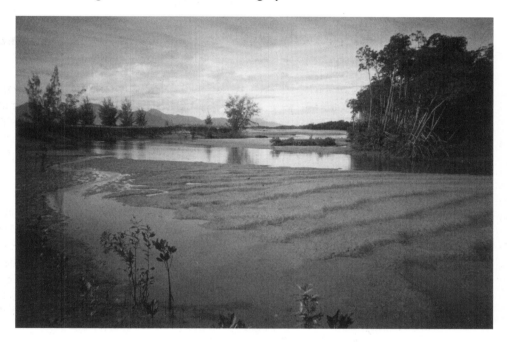

Plate 4.1 *A mangrove-lined tidally influenced channel, with sandy mega-ripples. A distributary channel of the Barron River Delta near Cairns, Queensland (Australia).*

Wave-dominated deltas

These occur where river-delivered sediment is redistributed by wave action. The onshore and alongshore sediment transport promoted by waves tends to produce rather linear or smooth shorelines (Plate 4.2). Where incident waves break parallel to the shore then the delta may develop symmetrically about the river outlet; however, where waves break obliquely to the shore then the delta morphology may be highly asymmetrical because longshore currents transport sediment downdrift. The São Francisco Delta of Brazil is an example of a wave-dominated delta.

Plate 4.2 *The linear shore of the wave-dominated Shoalhaven Delta, New South Wales (Australia).*

The Volta Delta in west Africa is asymmetrical and has recently become predominantly influenced by wave activity that has had far-reaching consequences for the delta and the downdrift coastline (Anthony and Blivi 1999). The delta has always suffered erosion of its shoreline by wave activity, which has transported this material downdrift along the coasts of Togo and Ghana, maintaining a barrier and lagoon system. Sand lost in this way by wave erosion was always replaced by river sediment derived from upland areas in the river catchment. However, in 1961 the Akosombo Dam was constructed on the River Volta which reduced riverine sediment input into the delta. Since then erosion by wave activity has dominated the delta shore and the barriers immediately downdrift of the river mouth (see also Box 4.1).

Box 4.1

The Danube Delta under wave attack

The Danube Delta on the Black Sea coast of eastern Europe is strongly affected by longshore currents. This drift is estimated to transport as much as 2,350,000 $m^3\ y^{-1}$ of sediment southward (Giosan *et al*. 1999). Shoreline changes have been measured for the period 1962 to 1987, indicating that extensive wave erosion is occurring along many stetches of the delta coast. This has been attributed to a reduction in sediment being delivered to the coast by the Danube River, and the construction of shore-protection structures, like groynes, that interrupt the longshore currents leaving local downdrift coastal sections deficient in sediment. A major barrier island is rapidly developing downdrift of the delta as a sink for the transported sediment.

Alternative delta classifications

Of course, within Galloway's (1975) model, two or more of the processes discussed above may be equally important so that some deltas may be river/tide-dominated tide/wave-dominated, or river/wave-dominated, some may be influenced more or less equally by all three. For example, the Niger Delta on the west African coast, possesses a concentric pattern of dominant processes, comprising an inner river-dominated zone, an outer wave-dominated zone, and a transitional tide-dominated zone. Similarly, the Red River Delta of Vietnam is influenced by the three processes, but manifested in sections, so that the section of the delta facing the fetch direction is wave-dominated, but the opposite side of the delta is sheltered and is dominated there by mesotidal conditions (Mathers and Zalasiewicz 1999).

Criticisms of Galloway's (1975) simple river–tide–wave-dominated model have been made in that it takes no account of, for example, sediment grain size and variations in sea level. The significance of grain size has been addressed by Postma (1990), for coarse gravel river-dominated deltas only, and Orton and Reading (1993) who have combined the range of grain sizes and processes to extend Galloway's ternary diagram into a prism (Figure 4.3). Gravel and mud/silt comprise

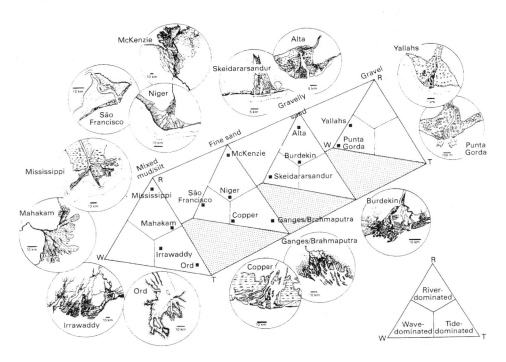

Figure 4.3 *Extension of the delta classification scheme outlined in Figure 4.2 to incorporate sediment particle size.*

Source: reproduced with permission from Reading and Collinson (1996) (Figure 6.2, p. 157), and with permission from the Houston Geological Society who are the publishers of Galloway (1975)

the two end members, with gravelly sand and fine sand as transitional categories between these end members. The influence of sea level change on delta development has been reviewed by Boyd *et al.* (1992), and based on earlier researchers they present a model showing that the regression or transgression of a delta shoreline is due to a combination of the rate of relative sea level change and the rate of sediment supply (see pp. 150–51).

Deltaic sediment supply

By definition, deltas are supplied with sediment from a fluvial source, and globally approximately 15×10^9 tonnes per year of sediment is transported into the sea by rivers. The type of sediment and the amount delivered to the coast are influenced by:

- the size and relief of the river catchment;
- the bedrock and its tectonic behaviour;
- the local climate, soil and vegetation; and
- the extent of human activity in the catchment.

Reading and Collinson (1996) suggest that large river catchments are associated with high volumes of fine-grained sediment (with coarser material being stored in the river system), lower gradients, and high and consistent discharges. Small catchments, on the other hand, are characteristic of tectonically active coasts, which are commonly uplifting with high gradients and energetic streams delivering coarse material to the coast. Climatic relationships are also evident, in that semi-arid

---

Box 4.2

Global significance of the Ganges–Brahmaputra Delta sediment flux

One of the largest deltas in terms of size and sediment flux is the Ganges–Brahmaputra Delta in the Bay of Bengal. Sediment gauges in the river system indicate an annual sediment discharge of 10^9 t y^{-1}. The delta has developed since sea level stabilised approximately 7,000 years ago, and comprises $1,500 \times 10^9$ m^3 of sediment in its subaerial part and $1,970 \times 10^9$ m^3 in its submarine part, which in this assessment includes the deep-sea Bengal Fan (Goodbred and Kuehl 1999). The significance of this immense sediment load lies in the fact that its source is the vast mountainous regions of the Himalayas and the Tibetan Plateau in central Asia. It is now widely accepted that the weathering of these mountains is consuming atmospheric carbon dioxide, and that river erosion transports this material, carbon and all, to the sea where it is locked away in marine sediments. It is this process that is now thought be a major influence in the initiation of global cooling and the onset of the Quaternary ice age (Raymo and Ruddiman 1992).

regions usually supply sand-grade sediment to the coast, whilst the humid tropics are more often associated with fine-grained silts and clays.

The main reason why the sediment delivered by rivers remains at the coast, enabling deltas to grow and develop, is due to shoaling and breaking of waves that favour the net landward transport of sediment. This hindrance to offshore sediment movement, with the exception of fine silts and clays, is known as the **littoral energy fence**. This fence may be overcome or bypassed under certain circumstances, along any coast, to allow sediment to move offshore, including:

- **river mouth bypassing** where rivers in flood jet out to sea due to unusually high discharge;
- **estuary mouth bypassing** where strong ebb-tidal currents draw sediment offshore; and
- **shoreface bypassing** where storm wave conditions create strong offshore currents close to the sea bed.

River discharge characteristics

The manner in which that sediment and water is discharged from a deltaic river into the sea is determined by a complex set of factors, such as the density difference between the fresh and saline waters of the river and sea respectively, the sediment type and concentration, water depths, water discharge volume, and river velocity (Reading and Collinson 1996). These in turn influence

- *inertia* of the river flow, that is the ability of the river water to continue moving into the sea before it is slowed or halted;
- *friction* between the flowing water and the bed and sides of the delta river channel and the sea bed beyond the river mouth; and
- *buoyancy* of the river water as determined by water density and/or water temperature differences.

Buoyancy is considered highly significant in delta development, and three buoyancy controlled flow regimes have been recognised, **homopycnal, hyperpycnal**, and **hypopycnal flow** (Figure 4.4).

Homopycnal flow

This occurs where the density of both the river water and the receiving waterbody are approximately equal. Sediment undergoes turbulent mixing at the river mouth and most deposition occurs at this location, especially of the bedload. Homopycnal flow typifies deltas that form in freshwater lakes, and is rare in the marine environment.

(a) homopycnal flow *(river = sea/lake water density)*

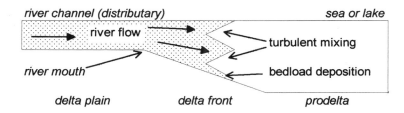

(b) hyperpycnal flow *(river > sea/lake water density)*

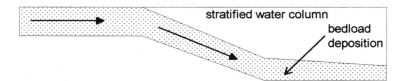

(c) hypopycnal flow *(river < sea/lake water density)*

Figure 4.4 *Three types of delta hydrodynamics based on the density differences between river and sea/lake water: (a) homopycnal; (b) hyperpycnal; and (c) hypopycnal flow.*

Hyperpycnal flow

This involves river water that is more dense than the receiving water body, so that the river water flows along the bed. Again, they are characteristic of deltas in freshwater lakes, where the inflowing river is colder than the lake water and often heavily laden with sediment. Stratification between the cold bottom flow and the warmer upper waters is most pronounced when the river enters deep water. River inflow into shallow water, however, forces a degree of mixing between the two layers which tends to obscure stratification.

Hypopycnal flow

This occurs where the river water is less dense than the receiving water body. This is commonly the case for rivers entering the sea, and hypopycnal flow typifies marine coastal deltas. Here, river water extends as a buoyant jet out into the sea, overlying the denser sea water. Furthermore, upon leaving the river mouth, the river water separates from the bed creating a salt wedge (similar to that discussed for estuaries on p. 91), and also separating bedload from suspended sediment. In this situation, coarse bedload is deposited at the tip of the salt wedge, where the river water becomes detached from the bed, and fine-grained suspended sediment is transported seaward within the buoyant plume of river water.

River mouth types

The influence of inertia, friction and buoyancy determine the configuration and morphology of deltaic river mouths. Three river mouth types are recognised:

1 **Inertia-dominated river mouths** form where a steep coastal slope occurs seaward of the river mouth, so that the river water is able to expand laterally as well as vertically, building up a uniform lobate delta morphology.
2 **Friction-dominated river mouths** are associated with very shallow water, so that considerable friction between river flow and sea bed forces the river flow to extend in a lateral direction only. This results in a generally long and narrow sediment build-up along the coast, which is dissected by outflow or tidal channels.
3 **Buoyancy-dominated river mouths** are characterised by hypopycnal flow, so that a relatively straight jet of river water extends out to sea, with little or no lateral extension along the coast. The bedload and the coarse suspended load are deposited at the river mouth as a distinct high bar, whilst the finer sediment is transported further out to sea before being deposited.

Box 4.3

The buoyant plume of the Rhône Delta, France

The Rhône Delta is located on the Mediterranean coast of southern France. The waters of the River Rhône enter the Mediterranean and extend far offshore as a readily identified buoyant plume. This plume has been mapped using radar techniques and has been studied through physical and geochemical methods (Broche *et al.* 1998). The results indicate that the morphology of the plume changes in terms of its orientation and in the distance reached offshore. Its orientation is largely determined by surface wind conditions, and the offshore limit largely reflects river discharge. The vertical structure of the plume in the water column also varies, from the classic two-layered situation, with the less dense plume overlying seawater, to a more mixed condition. Also, a complex multi-layered vertical structure has been observed. Studying the dynamics of deltaic freshwater plumes is important because of the exchange of material that takes place from the riverine to the marine system, and therefore, is significant in understanding water, sediment and geochemical budgets in the coastal zone.

River deltas

River deltas include all the major river deltas of the world, where generally large and low gradient rivers enter the sea (Table 4.2). The low to moderate energy of these rivers supply mud, silt and sand to the delta where it is reworked to varying degrees by wave and tidal processes (see pp. 111–14). Terminology used to describe delta sedimentology, morphology and processes are given in Table 4.3. The now classic sedimentary structure of deltas (Figure 4.5), first proposed by Gilbert (1885), but now known not to be universally applicable, comprises:

- subaerial horizontally deposited **topset beds**, laid down by the river and associated channels on the delta platform;
- **foreset beds** that are inclined draping the submerged seaward slope of the delta; and
- **bottomset beds** which are horizontally bedded deposits that occur seaward of the delta slope and laid down by the settling out of suspended sediment.

Deltas that clearly possess topset, foreset and bottomset beds have become known as **Gilbert-type deltas**.

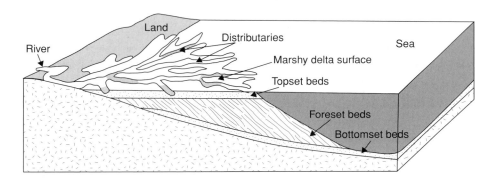

Figure 4.5 *General structure of a delta, indicating the location of topset, foreset and bottomset beds.*
Source: from Park (1997) (Figure 11.25, p. 325)

Of the terms given in Table 4.3, the process-based terminology is now widely used, and is a useful scheme for discussing the dynamic relationship between deltaic processes and morphology (Figure 4.6).

Delta plain

The delta plain comprises an extensive lowland environment, dominated by a river channel or a number of **distributary channels** that branch off the main river as it

Table 4.2 Examples of deltas and their river, sediment, delta plain and coastal environment characteristics.

Delta	River characteristics				Sediment	Delta plain		Coastal environment		
	Catchment (10³ km²)	Discharge (m³ s⁻¹)	Sediment load (10⁶ t y⁻¹)	Sediment/water ratio (g l⁻¹)	Grain size (mm)	Area (km²)	Gradient (m km⁻¹)	Wave height (m)	Tidal range (m)	Water depth (m)
Amazon	6,150	199,634	900	0.14	0.03	467,078	0.0125	moderate	4.9	100
Ebro	85.8	552	6.2	0.35	0.2	325	0.38		0.2	100
Fraser	234	3549	20	0.18	0.12–0.35	480			5	350
Ganges–Brahmaputra	1,597.2	30,769	1,670	1.76	0.16	105,641	0.05–0.17	low	3.6	
McKenzie	1,448	9,100	126		silty sand	13,000	0.05	low	0.2	70
Mekong	790	14,168	160	0.37	fine sand	93,781	0.02	low	2.6	
Mississippi	3,344	15,631	349	0.73	0.014	28,568	0.02	very low	0.4	
Niger	1,12.7	8,769	40	0.08	0.15	19,135		moderate	1.4	100–200
Nile	2,715.6	1,480	111	2.43	0.03	12,512	0.088	0.5–1.5	0.4	100
Orinoco	951.3	34,856	210	0.19	0.52	20,642	0.067	low	1.9	
Po	71.7	1,484	15	0.33	0.08–0.5	13,398	0.025–0.074		0.7	
Rhône	90	1,552	10	0.21		2,540				50–100
São Francisco	60.23	3,420	6			734		1	2.5	
Senegal	196.4	867.8				4,254		2.5	1.9	
Shoalhaven	7.25	57			0.25	85		1.5	1.2	
Yangtze	1,354.4	28,519	478	0.54	silt	66,669		1–1.5	2.8	50

Source: extracted with permission from Reading and Collinson (1996)

Table 4.3 *Comparison of terms describing the sediment structure, morphology and processes that operate in the three major deltaic areas (see also Figure 4.4).*

Sedimentological (Gilbert 1885)	Terminology type	
	Morphological (Nittrouer et al. 1986)	Process (Reading and Collinson 1996)
Topset beds	Delta platform	Delta plain
Foreset beds	Delta slope	Delta front
Bottomset beds	Prodelta	Prodelta

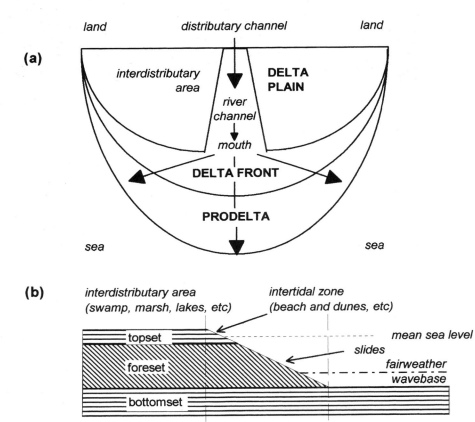

Figure 4.6 *Schematic geomorphology of a river delta: (a) plan view; and (b) shore-normal cross-section.*

Plate 4.3 *The extensive delta plain of the Shoalhaven Delta, New South Wales (Australia).*

approaches the sea (Plate 4.3). The channels are lined with levées (Plate 4.4), and in between the channels a number of different environments might occur, such as floodplains, lakes, salt or freshwater marshes, mangrove or freshwater swamps, tidal flats, sabkas, and marine embayments. Delta plains have been subdivided into upper and lower subzones (Reading and Collinson 1996), with the lower delta plain being characterised by the influence of tidal processes and the incursion of salt water, whereas the upper delta plain is dominated by fluvial processes. These subdivisions are perhaps not well named, as there may not be any significant altitudinal differences between the two subzones. Since their recognition is determined by the degree of marine inundation into the delta, terms such as outer and inner delta plain may be appropriate substitutes for lower and upper subzones respectively, and are used hereafter.

On a broad scale, then, delta plains are comprised of distributary channels and the areas between the channels, known as **interdistributary areas**. Channels on the inner delta plain behave similarly to fluvial channels. They normally have unidirectional flow and may meander according to the gradient. Channels on the outer delta plain however, behave more like estuaries (see pp. 86–97), in that flood tides penetrate the river mouths and may pond the river water, reducing water and sediment discharge, and in some cases where river mouth mixing is minimal, a distinct salt wedge may occur. Both fairweather and storm wave activity may also affect inner delta plain channels, which again may lead to ponding back of the river water. River ponding like this may result in flooding, and in some cases the switching of channels, from the old pre-flood channel to a new post-flood channel.

In addition to the riverine distributary channels, there may also be so-called tidally influenced channels that occur as shallow embayments or tidal inlets penetrating the outer edge of the delta plain. These are unconnected to the river system and may often be distinguished by an overall funnel-shaped form, narrowing inland, whereas river channels are usually parallel-sided.

The interdistributary areas can support a range of environments. Where they are vegetated, a zonation may be apparent, with the most landward areas being characterised by freshwater swamps that become brackish marshes or mangroves as the sea is approached. Areas between tidally influenced channels are usually intertidal, comprising salt marshes, mangroves or sabkas depending on the climatic setting, and in some cases dense tidal creek networks exist. Alternatively, interdistributary areas may be permanently or semi-permanently flooded to form shallow lakes or lagoons. A number of geomorphological features may occur in interdistributary areas, such as:

● abandoned channels and levées;
● **crevasse channels and splays**, which are formed where levées have been locally breached to produce a crevasse channel, and from which a crevasse splay may extend for short distances into the interdistributary area; and
● minor deltas, which may grow from an initially small crevasse splay and develop to such an extent so as to infill a low-lying interdistributary area.

Delta front

The delta front zone includes the river mouth, the shoreline of the delta, and the area immediately offshore. The geomorphological features encountered in this zone are very much dependent on the processes discussed on pp. 111–16 and 117–19. For example:

● River-dominated deltas tend to create river mouth sand bars that are orientated perpendicularly to the river flow, and also levées that may descend below sea level, defining a submarine channel.
● Tide-dominated deltas produce tidal channels and inlets separated by sand bars, that are usually aligned parallel to the tidal flow. Flood or ebb-tidal deltas may also be created within the river mouth.
● Wave-dominated deltas are characterised by beaches, spits and longshore bars, where wave activity has moved sediment onshore. Cheniers and aeolian dunes may also be present.

Of course, dominant delta processes may change through time and under different conditions. For example, under normal discharge conditions a river mouth may be buoyancy-dominated with a well-defined intruding salt wedge. However, with higher discharges, associated with storm rainfall for instance, the river mouth may become inertia- or friction-dominated. These changes will be reflected in the morphological features present at a given time (Reading and Collinson 1996).

Prodelta

This occurs in the form of an apron around the foot of the sloping delta front, lying in relatively deep water and characterised by fine sediment. The prodelta is below wave-base and so not affected by waves, and also tidal influences are minimal. Where the delta is in shallow water and the foot of the delta front is above wave-base, a fine-grained prodelta is unlikely to form. Prodelta sediment is deposited out of suspension from river water, outflowing into the sea as a buoyant plume. Laminations in prodelta sediments are common and represent regular fluctuations in the grain size of sediment discharge from the river, at times of different energy regimes, for example, summer low energy discharge and winter high energy discharge. These laminations are best preserved where the prodelta bottom waters are anoxic, otherwise the burrowing activity of benthonic organisms may destroy these subtle sedimentary textures (Reading and Collinson 1996).

Fan deltas

This is a collective term for a number of different delta types that are characteristically comprised of coarse-grained sediment, rather than the finer material of a river delta. The coarseness of the sediment is indicative of the relatively high fluvial energy involved in transporting the material to the coast. For fan deltas, the fluvial source is known as the **feeder system**, and Postma (1990) has recognised three distinct feeder systems:

1 Type A feeder system is a single high gradient river or stream channel, that runs off high relief areas, often tectonically uplifting. Local landslides as well as stream flood events deliver very coarse gravel sediment as bedload to the delta front and sometimes to the prodelta region.
2 Type B feeder system is characterised by a relatively high gradient, but this time a **braided river** (a single river course that comprises multiple channels) is involved. This type also occurs on tectonically active coasts, but is especially common in glacial environments, where the braided rivers are fed by glacial meltwater. In such cases, cold meltwater discharge may be very dense and the resulting hyperpycnal flow commonly transports coarse gravel material down the delta front to the prodelta.
3 Type C feeder system is characteristic of a moderate-gradient braided river or **braidplain** (a plain with several braided rivers), and often occurs in glacial environments. Despite the lower energy, however, sediment is still predominantly transported as bedload, but the sand content is usually higher than coarse gravels. Based on sedimentology, this feeder system type may be thought of as transitional between the true high gradient fan deltas and the low gradient river deltas discussed in the previous section (Reading and Collinson 1996).

Fan delta morphology

Fan deltas are very varied in their morphology as they often occur in tectonically active coasts, which may over time vary river gradients, and therefore energy inputs, and relative sea level. However, three broad types have been recognised (Figure 4.7) on the basis of gradient and water depth of the coastal shelf (Reading and Collinson 1996):

1 **shelf-type (shallow water) fan deltas** occur on low gradient and shallow water coastal shelves. The overall morphology is that of a gradually seaward sloping delta plain and front, extending gently down to the prodelta. The sediment is generally coarse, but a seaward decrease in grain size is often seen, and the

(a) shelf-type (shallow water) fan delta

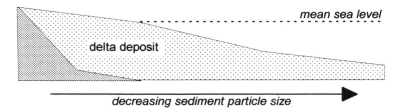

(b) slope-type (deep water) fan delta

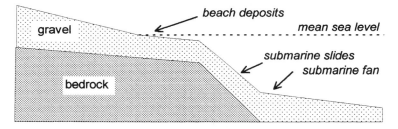

(c) Gilbert-type fan delta

Figure 4.7 *Schematic geomorphology of fan deltas: (a) shelf-type (shallow water) fan delta; (b) slope-type (deep water) fan delta; and (c) Gilbert-type fan delta.*

prodelta (if the water is not too shallow) is comprised of relatively fine-grained sediment. This represents a lateral zonation of sediment types.

2 **slope-type (deep water) fan deltas** occur in deep water and are characterised by a steep slope between the delta front and prodelta. In certain cases, there may be a submarine delta at the base of the steep slope, fed by submarine flows down the delta front. Generally, the overall morphology is characterised by a low to moderate gradient delta plain and a high gradient delta front. Lateral sediment zones include a gravelly delta plain that passes seaward into extensive beach deposits, that are perched at the top of the steep delta front.

3 **Gilbert-type fan deltas** are so called because they possess the distinctive topset, foreset and bottomset sediment layers that were first described by Gilbert (1885) (see p. 120). These fan deltas are steeply inclined throughout, from delta plain to prodelta, and may occur in both deep and shallow water. Mass movement processes are commonly active on these steep deltas, where slope gradients may be as much as 35°.

Because of the coarse-grained nature of fan deltas tides have little influence on sediment transport and reworking. Similarly, fairweather wave conditions play a limited role in this respect, but storm wave events and tsunami may be significant, particularly considering the high latitude and tectonically active locations where fan deltas are often encountered.

Holocene delta development

Sea levels during the Quaternary ice ages were much lower than today. Indeed, during the Last Glacial Maximum 18,000 years ago, sea levels were approximately 130 m lower than at present, and the world's shorelines were located close to the edge of the continental shelf. During the post-glacial Holocene Epoch, sea levels have risen, rapidly at first to about 6,000 years ago and then more slowly. Deltas that were active during the Last Glacial Maximum and formed at the continental shelf edge were either drowned by Holocene sea level rise, or survived if the rate of sediment supply, accretion and elevation of the delta surface always exceeded the rate of sea level rise. Where these deltas do survive they are termed **shelf edge deltas** (also known as shelf margin deltas) and they are often surrounded by relatively deep water. Where deltas did not elevate faster than sea level, then they migrated landward, always occurring in relatively shallow water and known as **inner shelf deltas** (also known as shoal water deltas). Most of the largest modern deltas, that protrude out on to continental shelves, are of this type (Reading and Collinson 1996).

Vertical sediment sequences in many modern deltas reflect the onshore migration of delta environments. A typical Holocene sedimentary sequence may comprise a basal layer of riverine sediments, such as floodplain deposits, overlain by delta plain distributary channel and interdistributary area deposits, passing upwards into tidal sediments, and finally shoreline deposits, which may include beach-dune sands. This sequence simply reflects the upward and landward migration of the various delta environments.

Such sedimentary responses to rising sea levels during the Holocene are experienced throughout the deltaic river system. Aslan and Autin (1999) report sedimentological changes through a 15–30 m thick Holocene sequence at a location 300 km upstream from the mouth of the Mississippi Delta. Here they find two sediment units:

1   a Lower Holocene unit, deposited prior to 5,000 years ago, consisting of swamp and shallow lake muds and deposits from numerous small stream channels – significantly there is little evidence for soil formation in the river environment at this time, and
2   an Upper Holocene unit, deposited from 5,000 years onwards, consisting of sand bodies representing river meanders, silts from levée formation, and evidence for soil formation.

Therefore, a dramatic change occurred 5,000 years ago in the depositional environment which has been interpreted as representing the deceleration of the rate of Holocene sea level rise. During rapid rise the Mississippi floodplain accreted rapidly, but slowed when sea levels stabilised following 6,000 years ago, which increased river meandering and allowed soils to form. Clearly, this location, presently 300 km inland, is affected by coastal processes and therefore, indicates

---

Box 4.4

Delta response to sea level stabilisation

Following the onshore migration of a delta and the stabilisation of Holocene sea level, a delta may prograde seaward if sediment supply is great enough. An example of this is the Shoalhaven Delta in New South Wales, Australia (Young *et al.* 1996). The Shoalhaven River is a major drainage system on the southeast Australian coast, with a substantial fluvial sediment load (Figure 4.8; Table 4.2). It appears that upon sea level stabilisation 6,000 years ago, the Shoalhaven River extended across a large embayment, as a levée-lined channel, to a wave-dominated sand barrier to which it supplied sediment; its descendant is now known as Seven Mile Beach. The deltaic back barrier area subsequently infilled with fluvial sediment, resulting in the seaward progradation of the shoreline by 3.5 km. The delta continued to be influenced by tides up until 2,500 years ago, but this influence declined as fluvial infilling progressed, and although the mouth of the Shoalhaven River is still tidal, tidal processes are no longer significant for much of the delta. Therefore, for many deltas rapid sea level rise resulted in landward retreat, whereas sea level stabilisation in conjunction with high sediment input leads to seaward delta progradation (see Plates 4.2–4).

Figure 4.8 *Shoalhaven Delta and its environments.*
Source: reproduced with permission from Young *et al.* (1996)

Plate 4.4 *Cows grazing on levées of the Shoalhaven River, part of the Shoalhaven Delta, New South Wales (Australia).*

how wide the coastal zone is if we strictly adhere to the definitions introduced on p. 3.

Deltas and human activity

Deltaic areas have long been attractive to human populations. This is due to a variety of reasons, including:

- *Topography* – deltas are usually of a uniform low relief, which makes communications routes within deltas relatively easy to establish, with the use of ferry boats, fords and bridges to cross waterways.
- *Fertility* – flooding of the delta plain supplies silt to the soil which renews its fertility. Therefore, deltas are excellent agricultural locations, and many countries have relied on harvests from deltaic areas for subsistence. This includes the Nile Delta which supported the Egyptian dynasties for thousands of years.
- *Reclamation* – new, highly fertile agricultural land can be brought into service through the construction of modest drainage systems in the wettest parts of the delta plain and front (e.g. swamps, marshes, shallow lakes), so increasing productive land.
- *Trade* – the rivers that create deltas are often large enough to be navigable to ocean-going craft, and therefore, major ports often arise in deltaic settings to import goods from abroad, and to export goods, including agricultural produce derived from the delta itself.

However, some of these attraction factors can also present problems. This is especially true of the floods involved in depositing silts. Bandyopadhyay (1997) lists a considerable number of natural hazards that are affecting Sagar Island (India) in the western part of the Ganges–Brahmaputra Delta. The island was reclaimed from mangrove wetlands from 1811 onwards, and is now fully populated. The effects of tropical cyclones, coastal erosion, tidal inundation and sand dune encroachment are just some of the hazards faced by the islanders. Addressing these hazards has led to seven different agencies having some involvement in protecting the island, often with little coordination. Protection schemes that have been initiated include the building of embankments, both along river courses and at the shoreline, replanting of mangroves and vegetation wind-breaks, resettlement areas for displaced people and refuges for when storms occur. The success of these schemes is variable.

Human impacts on deltas

Many of the human impacts already discussed in Chapters 2 and 3 apply equally to deltaic wave and tidal environments. However, impacts that are specific to deltas

Box 4.5

Cyclone hazards on the Ganges–Brahmaputra Delta

The Ganges–Brahmaputra Delta is situated in the Bay of Bengal, and spans three international boundaries, India, Bangladesh, and Myanmar (Burma). The Indian Ocean accounts for 10 per cent of the world's cyclones, with an average occurrence of 1.77 cyclones per year. These cyclones often track northwards into the Bay of Bengal, where the funnel-shape of the bay piles up the water at the delta coast, and in conjunction with storm surges and increased precipitation, wind velocities and wave heights, results in extensive flooding of the deltaic area (Figure 4.9).

Figure 4.9 *Areas of the Ganges–Brahmaputra Delta prone to flooding to depths greater than 90 cm.*
Source: from Pickering and Owen (1997) (Figure 8.7, p. 321)

Between 1797 and 1991 the area was hit by sixty severe cyclones, causing nearly one million deaths (Figure 4.10). The cyclone of 29 April 1991 was well documented and nearly 140,000 people died, most of whom were children under the age of ten, women

continued

and the elderly, and most drowned. Cyclone shelters had been constructed following the early 1970s (e.g. Figure 4.11), but these could accommodate only 450,000 of the 5 million people at risk (Park 1997). One million homes are estimated to have been destroyed and another million damaged. Agriculture was severely affected, with 60 per cent of cattle, 80 per cent of poultry, 113,300 ha of standing crops, and 72,000 ha of rice paddies destroyed. Coastal flooding was enhanced because some 470 km of protective embankments failed. Disease and famine affected the survivors, and the total economic impact of the event was US$2.4–4.0 billion (Kay and Alder, 1999).

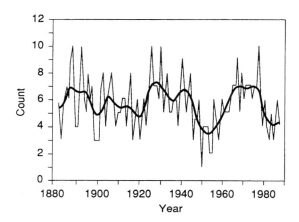

Figure 4.10 *Estimated number of tropical cyclones affecting the North Indian Ocean between 1880 and 1990.*
Source: Pickering and Owen (1997) (Figure 8.11b, p. 329)

Figure 4.11 *The cyclone track of 1970 which affected the Ganges–Brahmaputra Delta and killed approximately 300,000 people.*
Source: from Park (1997) (Figure 9.10, p. 269)

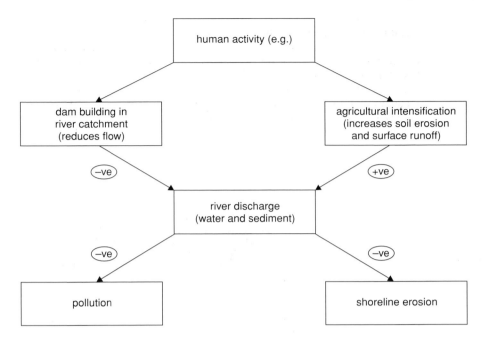

Figure 4.12 *The relationships and feedbacks between some examples of human activity and the dynamics of a delta system.*

have come to light since the 1970s, and originate from interference with the river systems that supply deltas with water and sediment (Figure 4.12). Interference includes:

- The expansion and intensification of agriculture within a river catchment, including deforestation, which may increase soil erosion and in turn the sediment supply to the delta, via the river system. This may lead to the silting up of distributary channels, which could increase the frequency and magnitude of delta plain flooding. Also, if the sediment is discharged from the river mouth, it may enhance shoreline sedimentation and the seaward progradation of the delta front.
- The construction of dams in the river system is known to severely affect delta dynamics. Both water and sediment discharges are reduced, which limits the capability of a river to flush itself out of sediment and pollution and reduces the sediment supply to the delta front, which often leads to shoreline erosion.

The latter impact is of concern in many major deltas worldwide, as many cities built on delta fronts are threatened by coastal erosion and inundation.

The Colorado River has had its flow controlled to the point of elimination for nearly a century, and the future of its delta is being determined by wave and tidal conditions in the Gulf of California. The reduction in freshwater discharge into the upper Gulf of California has forced a change from brackish water to hypersaline conditions, and sediment analysis shows that Colorado Delta sediments are currently being extensively reworked (Carriquiry and Sanchez 1999). Similarly, the water and sediment supply to the Luanhe River Delta in northern China has decreased rapidly since 1980, following water conservancy projects constructed in the river catchment. The Luanhe Delta only started to develop in its present position in 1915, due to the river switching its channel, and human interference is already leading to its abandonment (Feng and Zhang 1998).

Sediments along the shoreline of the Nile Delta in Egypt are being redistributed by wave activity, as a consequence of the construction of the Aswan Dam in 1964, and could have serious physical and socio-economic impacts, especially if sea levels rise in the near future (El Raey *et al.* 1999a, b). Such a future sea level rise (see pp. 153–160 for details) is predicted to have major impacts on human populations on many deltas. The Yangtze Delta in China is a peri-urban area around Shanghai with significant agricultural activity. Chen and Zong (1999) suggest that future sea level rise will result in erosion along the southern coast of the delta, and that sediment infilling in parts of the Yangtze Estuary will occur. It is argued that the risk of flooding will increase, that the water table will rise and induce prolonged waterlogging, and that during the dry season tidal incursions of sea water will reduce the amount of freshwater available for the irrigation of agricultural lands, which may lead to **salinisation**. Altogether, it is likely that the agricultural productivity of the region will be greatly reduced.

These examples show how sensitive delta systems are to human interference, and how extensive the system feedbacks are, so that actions taken far inland in the river catchment affect the environment and people living at the coast. Because of the magnitude of interference, deltas are the focus for many international monitoring studies, but because of the spatial scale involved, remote sensing techniques are widely employed, and are perhaps the only way of effectively recording erosion, accretion and general delta shoreline changes through time (e.g. White and El Asmar 1999).

Summary

- Deltas differ from other coastal environments in that their principal sediment source is derived from river systems, yet they are still influenced strongly by wave and tidal activity.

- Delta hydrodynamics are determined by the density differences between the river and sea water, which in turn largely determines the fate of river-borne sediment within the delta system.

- Delta geomorphology comprises a subaerial delta plain, a delta front and a submerged prodelta.

- The development of modern deltas has been strongly influenced by the post-glacial rise in sea level, which has caused deltas to retreat landward before prograding seaward after sea levels stabilised.

● Human impact on delta systems is varied, but of special concern is the interference of delta river system catchments, such as agricultural intensification and dam-building projects, which influence water and sediment discharges at the river mouth, often with feedbacks that alter shoreline positions.

Discussion questions

1 Assess the validity of using the contribution of wave, tide and river activity to classify deltas.

2 Examine the links between the hydrodynamics, sedimentology and geomorphology of delta systems.

3 Citing examples, evaluate the physical changes brought about through human interference in the river catchments of delta systems.

Further reading

See also

Tectonic classification of coasts, Chapter 1.
Longshore beach features on wave-dominated coasts, Chapter 2.
The role of tidal currents, Chapter 3.
Global warming and the threat of future sea level rise, Chapter 5.

General further reading

Clastic coasts. H. G. Reading and J. D. Collinson. 1996. In: H. G. Reading (ed.) *Sedimentary Environments: Processes, Facies and Stratigraphy*. Blackwell, Oxford, 154–231. A substantial chapter on modern and ancient clastic coasts, but with a clear emphasis on deltaic sedimentary environments.

Journal of Coastal Research, volume 14, part 3. 1998. A timely thematic issue devoted to deltaic coastal environments, with some excellent review articles, including the Mississippi Delta and Arctic deltas, as well as research papers. A good source of recent delta literature.

Coarse-grained Deltas. A. Colella and D. B. Prior (eds). 1990. Blackwell Scientific Publications, Oxford, 357pp. A compilation of research papers on fan deltas.

Deltas: Sites and Traps for Fossil Fuels. M. K. G. Whateley and K. T. Pickering (eds). 1989. Geological Society, London, Special Publication No. 41, 360pp. As the title suggests, this is a collection of geological research papers devoted to hydrocarbon exploration in deltaic regions.

5 Sea level and the changing land–sea interface

Sea level change is one of the main factors in stimulating coastal change, and in the long term it has been controlled by changes in the volume of landlocked ice sheets. If sea level rises, then coasts must be able to change dynamically to keep up, or drown. On the scale of a human lifetime we may consider sea level to be unchanging, but even into the twenty-first century it now appears that sea level has been rising due to human activity, and may continue to rise well into the future. This chapter covers:

- the principal mechanism for sea level change
- methods employed in constructing records of sea level change
- the various responses of coastlines to sea level change
- twentieth-century sea level rise, that has been attributed to our enhancement of the Greenhouse Effect, and future predictions
- ways of managing sea level rise

Introduction

The level of the sea is not constant, it is always rising and falling, either through the passing of waves (including tides), meteorological influences or gravitational effects in the form of the earth's **geoid**. Long-term and significant sea level changes, however, reflect changing levels of both land and sea; **eustasy** refers to absolute changes in global sea level, and **isostasy** refers to the vertical movement of land due to local geological factors. It is the balance between these two processes at a given coastline that produces observed changes in sea level, referred to as **relative sea level change**, because an absolute (eustatic) rise in sea level may not be required to allow the sea to rise relative to the land. Figure 5.1 explores this balance in a simplified way, and a number of cases can be put forward to illustrate the relationship:

- Relative sea level rise will occur if (a) eustatic sea level rises whilst the land is isostatically subsiding, static or uplifting at a slower rate than eustatic rise; (b) eustatic sea level is static whilst the land is isostatically subsiding; and (c) eustatic sea level falls whilst the land is isostatically subsiding at a faster rate than eustatic fall.
- Relative sea level fall will occur if (a) eustatic sea level falls whilst the land is uplifting, static or subsiding at a slower rate than eustatic fall; (b) eustatic sea

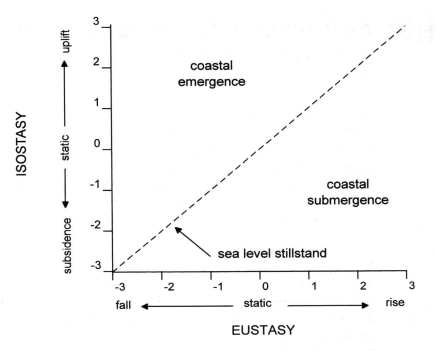

Figure 5.1 *The relationship between eustatic and isostatic changes and their influence on relative sea level and coastal emergence/submergence. Note: The numbers on the axes represent rates of change.*

level is static whilst the land is isostatically uplifting; and (c) eustatic sea level rises whilst the land is isostatically uplifting at a faster rate than eustatic rise.

- A sea level **stillstand** will occur if (a) eustatic sea level fall and isostatic subsidence of the land occur at the same rate; (b) eustatic sea level and the isostatic land level are both static; and (c) eustatic sea level rise and isostatic uplift of the land occur at the same rate.

The main reason for eustatic and isostatic changes in the level of the sea and land is the growth and decay of continental ice sheets during the Quaternary. Glacierization requires ice, which is introduced into glacial systems as snow, mainly derived from oceanic evaporation. Therefore, during Quaternary glacial stages water is extracted from the oceans and transferred to ice sheets where it is locked up for the duration of the glacial stage. This glacial control on eustasy is known as **glacio-eustasy**. Similarly, these same continental ice sheets growing on land exert great overburden pressure, causing the land to isostatically subside, a process known as **glacio-isostasy** (Figure 5.2). At the end of a glacial stage, melting ice returns water to the ocean causing sea levels to rise, and at the same time, the release of ice overburden results in the isostatic rebound or uplift of the land. Therefore, previously glaciated areas often possess very complex relative sea level histories, reflecting the balance between glacio-eustatic and glacio-isostatic changes (e.g. Gehrels *et al.* 1996;

(a) Simplified illustration of isostasy with wood blocks floating in water.

(b) Orogenic uplift, erosion, sediment transfer and isostatic adjustment.

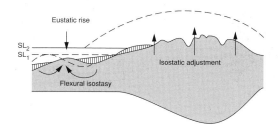

(c) Glaciation, crustal depression, complex margin responses and eustatic sea-level fall.

(d) Deglaciation, isostatic recovery, sediment transfer and eustatic sea-level rise.

Figure 5.2 *Two stages in glacio-isostasy from (c) ice-loading and crustal depression to (d) deglaciation and isostatic rebound (SL = sea level). Also, (a) shows a simplified isostatic model, and (b) isostatic effects from erosion and deposition.*
Source: from Briggs *et al.* (1997) (Figure 4.7, p. 54)

Dawson and Smith 1998). Furthermore, knowledge of these processes has led to the development of models that attempt to predict evolving coastlines since the last glacial through the construction of shoreline maps for different time periods (e.g. Lambeck 1995).

The Quaternary Period spans from 1.81 million years ago to the present day and comprises the **Pleistocene Epoch** (1.81 million years to 10,000 years ago) and the **Holocene Epoch** (10,000 years ago to the present). During the Quaternary sea levels fluctuated widely, with sea level highstands during the interglacials and lowstands during the glacial stages. Many interglacial sea level highstands attained or exceeded the present sea level altitude, and many fossil shorelines may be encountered above present sea level and appear to be 'raised', hence the term **'raised' beach** (Plate 5.1). The word 'raised' can be a misnomer, because many of these fossil Pleistocene shorelines are *in situ*.

In the British Isles, mapping isostatic readjustments following the last glacial at the end of the Pleistocene shows that northern Britain is currently uplifting by up to 2 mm y^{-1}, whilst southern Britain is subsiding by up to 2 mm y^{-1} (Figure 5.3). The two regions are separated by a fulcrum line that extends in a stable zone from the

Figure 5.3 *Isostatic map of Great Britain: (a) showing areas of isostatic uplift (grey shading) and those threatened by sea level rise (black shading); and (b) rates of isostatic movement in mm y^{-1}.*
Source: from Briggs *et al.* (1997) (Figure 17.20, p. 320)

Plate 5.1 *A fossil Pleistocene 'raised beach' at Brigneoc'h, Brittany (France). Note the fining up sequence from large boulders to small pebbles. The cliff is approximately 3 m high.*

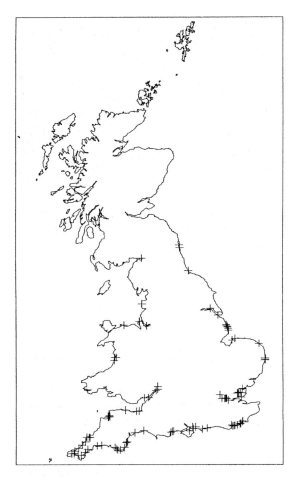

Figure 5.4 *The distribution of submerged forests in Great Britain indicating main areas of isostatic subsidence.*

Tees Estuary in northeast England, through the Dee Estuary and on to the Lleyn Peninsula in northwest Wales. This post-glacial isostatic movement is reflected in the distribution of **submerged forest** sites in Great Britain (Figure 5.4). The majority of sites are located in southern Britain, whilst only a few are found north of the fulcrum line. This indicates that post-glacial eustatic rise coupled with isostatic subsidence in southern Britain has led to relative sea level rise and the drowning of previously wooded landscapes, whilst in northern Britain isostatic rebound is outpacing eustatic rise, so that relative sea levels are actually falling there and resulting in coastal emergence rather than submergence.

Only in isostatically stable areas of the world can records of eustatic or absolute sea level change be possibly found, i.e. areas that have never been glaciated or suffer earthquakes, volcanic or other tectonic activity (earth movements) that may vertically displace the land. Such areas are rare, but the island of Bermuda is considered to be a stable mid-ocean carbonate platform from which a eustatic sea level record has been constructed (see Lowe and Walker 1997). Eustatic sea level curves from such areas indicate that the last glacial sea level 18,000 years ago was approximately 130 m below the present sea level. A period of very rapid rise ensued until approximately 6,000 years ago, but since then sea level has been comparatively stable with a less than 10 m rise to the present day. Figure 5.5 shows sea-level rise from 18,000 years ago to the present, and how the coastline of Great Britain has developed.

Constructing records of sea level change

Mean Sea Level is known to have varied altitudinally through time. Ample evidence exists to suggest that major changes in sea level have occurred through geological time. Evidence occurs in the form of:

(a) (b) (c) (d)

Figure 5.5 *Sea level rise and the development of the coastline of the British Isles: (a) sea level rise since 18,000 years ago; (b) extent of the last ice sheet; (c) early Holocene coastline; and (d) late Holocene coastline (broken line indicates position of recently abandoned shoreline).*
Source: from Briggs *et al.* (1997) (Figure 1, p. 308)

- **Erosional landforms**, such as cliffs, shore platforms, marine notches, sea caves, and arches which now occur well above present sea level or indeed are now submerged.
- **Depositional features**, such as fossil beaches, tidal flats, and coral reefs, again either above or below present sea level.
- **Biological indicators** i.e. organisms which when alive live at known tidal levels, but become stranded by changes in sea level, so that their occurrence as fossils indicates the position of the old shoreline. Examples include barnacles, tubeworms, and oysters.
- **Archaeological remains** can sometimes indicate sea level change, such as submerged dwellings or ports that have been elevated above sea level. Greece is well endowed with such remains, but also now submerged megaliths (standing stones) in Brittany (France) are excellent examples.
- **Historical documents and records** can indicate recent changes, this includes tide-gauge records that are available for in excess of 100 years at some locations, such as Brest in northwest France.

Changes in sea level through time are most frequently depicted as linear curves, illustrating the altitude of sea level for particular times in the past. These curves should be constructed using **Sea Level Index Points (SLIPs)** which can be any of the sea level indicators discussed above (pp. 138–140) for which age, altitude, **tendency of sea level**, and **indicative meaning and range** are known (see below for explanations). In the past, sea level curves have been produced with reference to age and altitude only, but these curves indicate very general sea level trends only and are difficult to compare from region to region, so care should be taken when consulting early sea level curves.

Dating sea level index points

Classical archaeological remains and historical documents quite often have good dates assigned to them, based on the period from which they originate, but for natural sea level indicators, laboratory and field dating methods must be employed. The technique chosen for dating a sea level index point depends to a great extent on the sea level indicator under investigation. Erosional landforms are notoriously difficult to date in absolute terms, that is, to assign an age in years. However, Haslett and Curr (1998) have attempted, for example, to date the relative ages (i.e. indicating which one is older than another without establishing actual dates) of a number of Quaternary erosional shore platforms in Brittany (France) using the weathering characteristics of the local granite bedrock. This works on the assumption that the oldest platform should be more weathered than platforms created by subsequent sea levels, because they have had more time to weather. Although the results are not straightforward, the technique demonstrates its potential as a useful discrimator of relative age. Erosional features are more easily dated if they have biological remains attached to them, such as fossil barnacles still attached to abraded surfaces, as discussed below.

For Late Quaternary sea level changes by far and above the most frequently used technique is that of the **radiocarbon dating** method. Radiocarbon (^{14}C) naturally occurs in the atmosphere and is incorporated into all living organisms at the same concentration as occurs in the atmosphere. Upon death ^{14}C undergoes radioactive decay, and radiocarbon dating uses the decay of this carbon isotope to date carbonaceous material, including preservable material such as wood, bone, peat, and shell. The half-life of ^{14}C was calculated in the 1940s as being 5,568 years, and therefore by measuring how much ^{14}C is left in a sample it is possible to estimate time elapsed since death. It is now known that the half-life of ^{14}C is in fact 5,730 years, but the initial estimate is still used so that comparison with early radiocarbon dates can be made. The measurement of ^{14}C in a sample is achieved either by the conventional technique of counting radioactive particles emitted from a sample, or by directly measuring the ^{14}C atoms in a sample using accelerator mass spectrometry (AMS) (Mannion 1999). Because the AMS technique makes a direct measurement, rather than relying on radioactive emissions, it is more accurate and can be performed on very small samples.

Although radiocarbon dating has become a widely used technique, it is not without uncertainty. Importantly, tree-ring studies (dendrochronology) have shown that radiocarbon years are not the same as calendar years. Beyond 2,000 years ago it appears that radiocarbon dates yield consistently younger ages than they should; for example, a radiocarbon age of 6,000 years is in fact closer to 7,000 calendar years old. Therefore, radiocarbon years have been calibrated by tree-ring studies, and all published radiocarbon dates are now calibrated and given as calendar years before present (cal. yrs BP) (e.g. Stuiver and Reimer 1993). Care in comparing radiocarbon dates in the older literature with more recent publications is therefore advised. Also, all radiocarbon dates possess error margins, and are always supplied with a ± figure. At the 1σ level (σ is sigma, and 1σ = 1 standard deviation) there is a 68 per cent probability that the age of the sample falls within the date provided, and at the 2σ level (2 standard deviations) there is a 95 per cent probability that the date is correct.

The selection of a sample for radiocarbon dating requires in-depth consideration. First, there is the issue of contamination to consider, for example, have the roots from younger plants grown down into the deposit under investigation, or has hard groundwater precipitated older carbon in the form of calcium carbonate in the pores of the sediment to be dated? Second, the derivation of the carbonaceous material should be considered. A peat may be considered **autochthonous**, that is occurring *in situ* and providing an accurate date for the deposit, whereas organic fragments in a sand deposit may be **allochthonous**, that is material transported by currents before being deposited, and therefore the carbon may be much older than the deposit in which it finally becomes incorporated. Considering the reasonably high cost of radiocarbon dating, all these issues need to be addressed.

Because of the relatively rapid decay of ^{14}C, radiocarbon dating cannot be performed on material older than about 60,000 years, and indeed material more than 20,000 years old is susceptible to yielding erroneous dates. Therefore, in older material alternative dating methods are required. **Uranium-series dating** is similar to radiocarbon dating in that it utilises the radioactive decay of uranium, thorium and radon isotopes to a stable lead daughter isotope. The decay process is considerably longer, so that older uranium-bearing deposits can be dated, including carbonate material such as coral and sea cave deposits (speleothems) (Mannion 1999).

A now common technique for relatively dating older Quaternary coastal sediments is that of **Amino Acid Racemisation (AAR)** (Lowe and Walker 1997). This requires fossil mollusc shell material of the sort commonly encountered in fossil beach deposits, such as mussels, cockles, and winkles, from which amino acids can be obtained. Amino acids persist in shell long after death, but convert through time from L-isoleucine to D-alloisoleucine after death, i.e. the process of racemisation. Therefore, the L/D ratio indicates time since death. The now seminal work of Davies (1983), on fossil beaches of the Gower Peninsula (Wales, UK), showed that the AAR technique could distinguish between beach remains deposited during different Quaternary interglacial stages. It must be remembered, however, that this is a relative dating tool only and relies on other methods to assign absolute ages to the L/D ratios.

Dating samples in the absence of organic remains can also now be achieved through a number of techniques, with perhaps the most widely used in coastal studies being **thermoluminescence (TL) dating** (e.g. Bryant *et al.* 1997). Crudely, this works because crystal sediment particles, such as quartz sand, release thermoluminescent energy when exposed to sunlight (i.e. when lying on a beach or in dunes), but when these crystal particles become buried they begin to store this energy. Therefore, by measuring the amount of energy accumulated in a buried sediment it is possible to estimate when it was last exposed to sunlight. There are some potential sources of error with this technique, but generally it works well and dates derived from this technique correspond well with dates derived from other techniques, such as radiocarbon dating (see Bryant and Price 1997, for further discussion).

The altitude of sea level index points

Establishing the altitude of a sea level index point requires accurate surveying using a theodolite, and altitude should be given relative to the national survey datum used in the country in question. For example, in the United Kingdom, the survey datum is known as Ordnance Datum (OD) (Newlyn) (see pp. 77–78 for explanation), and all UK altitudes are given relative to OD. Surveying erosional landforms is fairly straightforward, as long as the surveyed feature is well defined, such as a wave-cut notch or shore platform. The altitudinal range of the feature should be established, such as the top and bottom of a wave-cut notch. This requirement results in vertical features, such as cliffs, sea caves and arches, appearing to have large altitudinal ranges and so reducing the usefulness of these data in accurately reconstructing sea level changes.

Sea level index points based on depositional evidence are not only more readily dated than erosional landforms, as discussed in the previous section, but levels within deposits to be surveyed are often very restricted as regards their altitudinal range. For example, the sediment boundary (contact) between marine and terrestrial deposits commonly makes excellent sea level index points, and these contacts are often sharp and well defined so that they can be surveyed down to the millimetre scale if required. In coastal lowlands, sediment coring techniques are often employed to survey the buried sediments and in such cases the altitude of the ground surface is established, from which the depth to the sediment contact is subtracted to give the altitude of the sea level index point.

Some care is required in many cases in assigning altitudes to depositional sea level index points. This is particularly true where **isolation basins** are concerned (e.g. Long *et al.* 1999). Isolation basins are bedrock basins that at one time were submerged or at least inundated by tides, but have since been isolated from the sea by a relative fall in sea level. The altitude of the boundary between the marine deposits, that occur in the bottom of isolation basins, and overlying freshwater/terrestrial deposits, does not necessarily represent the altitude of a former sea level. This is because it is the altitude of the rock lip or sill of the bedrock basin that determines the sea level altitude that is capable of inundating the basin. Thus, an isolation basin may only be completely isolated from the sea when the level of highest astronomical tide falls below the altitude of the rock sill;

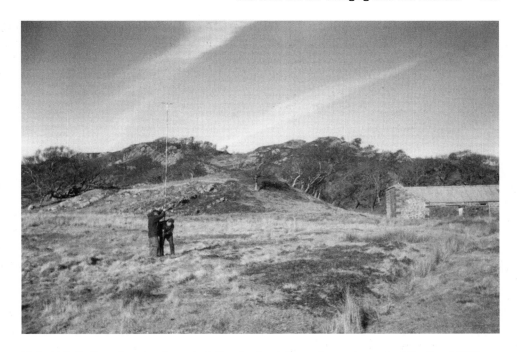

Plate 5.2 *Sediment coring in an isolation basin at Rumach, west coast of Scotland (UK).*

however, the upper surface of the marine sediments in the basin may be several metres lower than the rock sill altitude (Plate 5.2).

Further altitudinal complications that may arise with depositional sea level index points include the effect of sediment compaction. When soft sediment is buried by subsequent deposition, the overburden weight of this additional sediment causes compaction and therefore, a lowering of altitude from that at which it was originally deposited (Allen 1999). Compaction, if undetected, can introduce significant altitude errors into a sea level curve, and therefore, measures must be taken to estimate the amount of compaction that may have occurred (see Box 5.1). Not all sediment types compact to the same degree, indeed, Allen (1999) suggests that clay, silt and sand deposits suffer minimal post-depositional compaction, but that spongy organic deposits such as peat may be greatly compacted.

Tendency of sea level

This simply refers to whether sea level was rising or falling at a sea level index point and is most easily applied to deposits. This is established with reference to sediment directly underlying the sea level index point, so that if sea level has appeared to rise from the underlying sediment to the sea level index point (e.g. a transition from freshwater peat to marine clay), then a positive tendency is assigned.

Box 5.1

Sediment compaction and sea level curves

For sea level index points that involve peat it is essential that any post-depositional compaction is established so that the correct altitude can be assigned. In the Somerset Levels, an extensive coastal lowland in southwest England (Plate 5.3), Haslett *et al.* (1998a) describe a Holocene freshwater peat deposit overlain by marine clay (Figure 5.6). The peat blankets undulating bedrock topography with an altitudinal range of the upper peat surface spanning 2.22 m. Radiocarbon dating demonstrates the peat surface to be the same age along the length of the investigated transect, indicating that either over 2 m of 'instantaneous' sea level rise had occurred or that the peat surface was originally horizontal and subsequently deformed by compaction. Palaeoenvironmental analysis of the peat to clay contact indicates that the transition from freshwater peat to intertidal salt marsh was gradual and not 'instantaneous', further suggesting that peat compaction has taken place. In this case, 2.22 m of peat compaction occurred with a maximum clay overburden of just 3.16 m. This is worrying as many previous sea level curves for the Somerset Levels and elsewhere are based on Sea Level Index Points at peat-clay contacts of 20 m or more depth. Therefore, many early sea level studies need to be reviewed, and more realistic allowances made for sediment compaction, especially where peat deposits are involved.

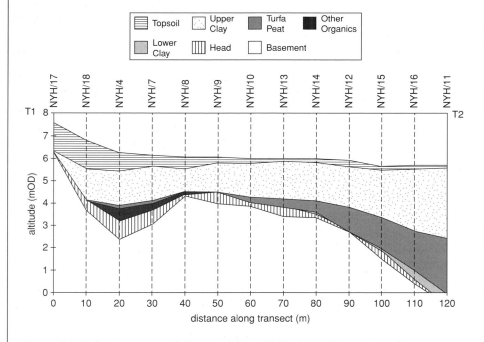

Figure 5.6 *Holocene peat and clay overlying solid bedrock. The peat surface was originally horizontal but has been compacted by the deposition of the clay. If unrecognised, compaction can introduce significant errors into sea level curves.*
Source: updated from Haslett *et al.* (1998a)

Plate 5.3 *The Somerset Levels – an extensive low-lying coastal wetland in southwest Britain, underlain by a thick sequence of Pleistocene and Holocene estuarine and marsh deposits.*

A negative tendency applies to where sea level appears to have fallen from the underlying sediment to the sea level index point (e.g. a transition from marine clay to freshwater peat). Sea level curves constructed without indicating tendency simply show the general trend of sea level change, whereas when tendency is indicated (by convention with arrows on each point) oscillations in the general sea level trend can be identified, which may be important in interpreting coastal evolution.

Indicative meaning and range

Indicative meaning refers to the position occupied by a sea level index point, when it was formed, relative to a reference water level. Usually the reference water level is mean sea level, so that indicative meaning refers to the position of a sea level index point within the tidal frame at the time. For example, a fossil salt marsh sediment surface may be suitable as a sea level index point, and may have been deposited at mean high water springs (MHWS); therefore, the indicative meaning of this sea level index point is MHWS. Indicative range refers to any errors that may be involved in assigning an indicative meaning, so that if it is not certain at what level a sediment was deposited between MHWS and highest astronomical tide (HAT), then the indicative range would be MHWST–HAT.

Establishing the indicative meaning of a potential sea level index point can really only be achieved through using biological remains (fossils). Many organisms live at, and are restricted to, particular tidal levels. Recovering these tidally-specific organisms either from sediments or attached to erosional rock surfaces can often tell us quite precisely the indicative meaning of a deposit or landform. Analysis of diatoms, foraminifera (Box 5.2), amoebae, ostracods, molluscs, crustaceans, pollen and other plant remains have been used with some degree of success in determining indicative meaning (see Tables 5.1 and 5.2).

Table 5.1 *Biological remains which are useful in determining the indicative meaning and range of sea level index points, with references for examples.*

Type of biological remain	Reference(s)
Diatoms	Zong (1998)
Foraminifera	Scott and Medioli (1978), Gehrels (1994), Haslett *et al.*, (1998b)
Ostracods	Boomer (1998)
Testate amoeba	Charman *et al.* (1998)
Molluscs and crustacea	Laborel and Laborel-Deguen (1994), Laborel *et al.* (1994)
Pollen	Shennan *et al.* (1994)
Other plant material	Gehrels (1994)

The main purpose for obtaining information regarding indicative meaning and range is to enable a comparison to be made between different sea level index points and to construct meaningful sea level curves. If a group of sea level index points for a particular region possesses different indicative meanings then they must all be standardised before a comparison and sea level curve can be made. Standardisation is achieved by reducing the points to Mean Sea Level, which is done by subtracting the modern tide level altitude of the coast in question from the altitude of the sea level index point. For

Table 5.2 *The distribution of biological structures in relation to sea level on rocky coasts.*

Tide level	Zone	Environment	Characteristic structures/species
High tide	Supralittoral zone (also known as the littoral fringe)	Rarely submerged, but wetted by spray	Lichens and blue-green algae (cyanobacteria)
	Midlittoral zone (also known as eulittoral zone)	Regularly submerged by waves and tides	Barnacles, algal ridges parallel with the sea surface, hard rim-like accretions of the coraline algae *Lithophyllum lichenoides* and notch erosion by cyanobacteria and *Patella*
Mean tide			
Low tide	Sublittoral zone (also known as infralittoral zone)	Mostly submerged, but upper part exposed at low tide	Reef-life constructions built up by vermetid gastropods and coraline algae, but rock is bored by clionid sponges, sea-urchins (*Paracentrotus*) and the bivalve *Lithophaga lithophaga*

Source: from Laborel *et al.* (1994); Pirazzoli (1996)

Box 5.2

Foraminifera and their use in sea level and coastal studies

The application of foraminifera to sea level studies is well established (e.g. Scott and Medioli 1978, 1986). Salt marsh foraminifera have been shown to occur within specific vertical zones on modern salt marshes from eastern North America (e.g. Gehrels 1994) which can be related to tide levels and accurately employed as tide level indicators in palaeo-salt marsh sediments. Of greatest significance is the recognition that in micro- and meso-tidal salt marshes of eastern North America, foraminifera extend up to highest high water, with the highest zone (Zone 1A) characterised by a monospecific assemblage of *Jadammina macrescens*, which extends down to mean highest high water. The recognition of Zone 1A in palaeo-salt marsh sediments is thus a very useful indication of palaeo-tidal level. Haslett *et al.* (1998b), and subsequently Horton *et al.* (1999), have investigated modern salt marshes in the United Kingdom and established foraminifera zonations, which differ slightly from their North American counterparts (Figure 5.7). Foraminifera are also employed as indicators of coastal pollution (e.g. Alve 1995) and sediment transport (e.g. Haslett *et al.* 2000a).

Tide Levels	Altitude (mOD)[1]	*Jadammina macrescens*	*Trochammina inflata*	*Ammonia beccarii*	*Haynesina germanica*	Local Zonation	Equivalent North American Zone[2]	Local Environment
	7.5					barren zone	High Marsh Zone 1A	High Marsh
MHWS	7							
	6.5					*J. macrescens* and *T. inflata* zone	High-Middle Marsh Zone 1B	Middle Marsh
	6							
	5.5							
	5					*A. beccarii* zone		
	4.5							
	4						no direct equivalent zones	Low Marsh
MHWN	3.7					*H. germanica* zone		
	3.5							
	3							Mudflat
	2.5							

[1] altitude of tidal levels refers to Oldbury-on-Severn
[2] North American zonation based on Scott and Medioli (1986)

Figure 5.7 *The relationship of foraminifera distribution to tidal levels and salt marsh environments in the Severn Estuary (UK). The restricted altitudinal ranges make foraminifera excellent tools for reconstructing past sea levels.*
Source: updated from Haslett *et al.* (1998b)

example, take a British macrotidal coast where MHWS is currently 6 mOD and MHWN is 5 mOD, and in coastal lowlands adjacent to the present coast two potential sea level index points have been established by sediment coring, one with an indicative meaning of MHWS at an altitude of 5 mOD and another with an indicative meaning of MHWN at 4 mOD, both suggesting a positive sea level tendency with a radiocarbon age of 4,000 cal. yrs BP. Although there is a 1 m altitude difference between these two points in the sediment cores, when reduced to mean sea level both points indicate mean sea level to have been −1 mOD at 4,000 cal. yrs BP (i.e. 5 − 6 = −1 mOD, and 4 − 5 = −1 mOD). If the indicative meaning for these points were not known, then it may appear misleadingly that an 'instantaneous' 1 m rise in sea level had occurred.

Coastal responses to sea level change

Relative sea level fall and the retreat of the sea from the land are known as a **regression**, whilst relative sea level rise and the associated inundation of coastal land are known as a **transgression**. Regressions and transgressions lead to the emergence and submergence of coasts respectively. The manner in which a given coastline responds to such sea level changes depends on the character of the coast, that is whether it is dominated by depositional or erosional processes. We discussed earlier the relationship between eustasy and isostasy in determining relative sea level (Figure 5.1), however, along depositional coasts sediment can act as a substitute for, or in addition to, isostasy in certain circumstances. If a beach has an abundant sediment supply, then it may build upwards and advance into the sea regardless of the eustatic regime, a process known as **progradation**; oppositely, if sediment supply is limited and less than that lost from or redistributed within the beach system, through erosion and the action of currents for example, then a beach may retreat landward (i.e. **retrogradation**) under a rising relative sea level, although still maintain its relative tidal position if it is allowed to translate landward (Swift 1975). Bruun (1962) suggested a model that describes the response of a sandy coast to sea level rise, where sea level rise forces the beach to retreat landward through the erosion of the back beach area. This eroded material is then redeposited at the foot of the beach, so that the beach may maintain its equilibrium profile. This model, commonly referred to as the **Bruun rule**, has been widely applied and yet has a number of unresolved problems, such as how can we know when a shoreline is in equilibrium, what timescales are involved in morphological response, and is it fairweather or storm conditions that drive the retreat process (Healy 1991)?

The response of different types of coastal systems to sea level change is often very specific to the given system. Relative sea level fall usually leads to the emergence, abandonment and fossilisation of a shoreline, and indeed many fossil Pleistocene shorelines are created in this way, such as 'raised' beaches. However, sea level rise often leads to the landward migration of a mobile shoreline or the submergence and drowning of a stationary shoreline. The response of different coastal systems to sea level rise during the Holocene and into the future has received much attention, particularly for the following examples of vulnerable and dynamic soft sediment coastal types.

Salt marshes

The response of salt marshes to sea level rise depends to a very great extent on the sediment supply into the salt marsh system. The balance between sea level rise and salt marsh surface elevation, through vertical sediment accretion, is critical (Reed 1995). If the rate of accretion and surface elevation is less than the rate of sea level rise then the salt marsh will submerge and drown; if the rate of elevation is higher than sea level the salt marsh will emerge and perhaps terrestrialise; the rate of surface elevation may be equal to that of sea level rise, and in which case the salt marsh will neither emerge nor submerge and therefore, exist in equilibrium.

Mangroves

Mangroves were discussed in general on pp. 106–108, but the response to sea level rise in Northern Australia has been examined by Woodroffe (1995). He suggests that mangroves will respond in different ways according to their overall morphology and physical setting. In the tide-dominated Darwin Harbour, mangroves occur fringing the estuary, with many tidal creeks supplying the system with sediment. Therefore, under gradual sea level rise, this fringing type of mangrove may migrate landward, supplied with sediment from erosion of the submerging lower intertidal zone. Palaeoecological studies suggest that all the studied mangroves migrated onshore this way during the post-glacial Holocene transgression. However, in other areas that are river-dominated, such as the South Alligator River, extensive deltaic plains of mangroves have been built as the shore has prograded seaward, following the stabilisation of the Holocene transgression, using sediment delivered to the coast by rivers. These near-horizontal plains are susceptible to any future sea level rise, because they are less likely to be able to migrate landward, and may simply become submerged.

Coastal sand dunes

Foredunes become scarped (Carter and Stone 1989) as they are reached by the rising sea level, and this releases sediment back into the beach system, perhaps for use in dune building further inland. Sea level rise may also breach the foredunes to deposit sand behind the foredunes as wash-over fans. Through both scarping and breaching, sea level rise promotes the landward migration of coastal dunes. Along many coasts during the Holocene transgression, sand dunes were blown inland ahead of the migrating shoreline. Dune sand blown on to the top of low cliffs and now perched there, offer an opportunity to date this initial sand inundation along a coastline (e.g. Haslett *et al.* 2000b).

Gravel beaches

Current research on the effect of sea level rise has suggested that gravel shorelines may respond as follows:

- The crest may build up in height by a transfer of material to the crest from the mid-tide zone as sea level rises. A barrier may start to migrate landward through the process of **rollover**, which is the general transfer of sediment to the landward side of a barrier (Plate 5.4).
- In the absence of adequate sediment supply, the beach will steepen leading to increased reflectivity.
- Crest build-up cannot continue indefinitely, as crest construction requires spilling waves for sediment transport, which in turn requires a measure of dissipative

Plate 5.4 *Barrier rollover at Porlock (Somerset, UK): here pebbles can clearly be seen encroaching onto agricultural land behind the barrier.*

character, so as reflectivity increases the rate of crest build-up will slow, and this shift in morphodynamic domain from dissipative to reflective will induce cusp formation.

- Cusps reaching the crest may then act as a template for overwash and breaching, and the eventual break-down of a barrier.
- In this way barriers will tend to migrate landward under a sea-level rise regime. Beaches confined by backing cliffs may develop swash ramps extending several metres above the high water mark, and with cusp development, ramps may become fragmented and separated by intra-beach erosional bays.

A characteristic feature of a landward migrating barrier is the presence of a terrace or terraces in the seaward low- to mid-tide zone. These terraces are low-angled and dissipative. Erosional slopes are often formed as a barrier migrates landward. Where a rock platform is exposed, it may supply the barrier with clasts thus slowing down migration, so the areal ratio of barrier to terrace can increase as well as decrease. There is also a type of depositional terrace resulting from sea-level rise and landward barrier migration, which is known as a boulder frame. This frame results from particle sorting, providing a residue or lag of larger and relatively immobile clasts at the foot of beach. If this residue of clasts remains immobile, then as the barrier migrates landward, the frame will expand in width. With the expansion of low-angled terraces, within the barrier morphodynamic system, reflectivity will change from being the dominant domain, to co-dominant to non-dominant through time, giving rise to an overall dissipative domain, which may lead to barrier abandonment. Furthermore, where a barrier progressively loses material to its frame as it migrates, the barrier's migration rate may increase, which may ultimately overstretch the barrier, leading to breaching. In order to maintain stability, therefore, the rate of volumetric loss needs to be balanced by the increasingly dissipative role of the frame (Carter and Orford 1993).

Global warming and the threat of future sea level rise

Global temperatures rose by approximately 0.6°C during the twentieth century and climate models estimate that this figure is set to rise to 2°C by 2100 (Houghton *et al.* 1996). This global warming has been attributed in part to human activity, and in particular to the burning of fossil fuels that release carbon dioxide (CO_2) into the atmosphere. CO_2, methane (CH_4), chlorofluorocarbons (CFCs), tropospheric (low-level) ozone (O_3), water vapour (H_2O) and some other compounds, are important gases that are able, in the atmosphere, to absorb heat radiated by the earth, whilst allowing the sun's energy to pass through unhindered. Therefore, these gases allow the atmosphere to act like a greenhouse, and are responsible for producing the earth's average temperature of 15°C. This has given rise to the phenomenon being referred to as the Greenhouse Effect, and without this natural phenomenon, the earth's average temperature would be in the region of −17°C. Concern is focused

> ## Box 5.3
>
> ### Sea level rise, coastal retreat and ecosystem impacts
>
> The impact of sea level rise on coastal habitats and wildlife is reviewed by Biljsma (1996); however, one major ecosystem impact that will affect developed countries is the phenomenon known as '**coastal squeeze**'. This involves coastal systems which respond to sea level rise by migrating onshore, but are unable to do so because of coastal protection constructions or natural barriers. Pethick (1993) documents such a case from eastern England, an area that is isostatically subsiding, where he recognises extensive estuarine mudflats and salt marshes that are migrating up estuary as sea level rises. The estuary is lined by sea walls which act as barriers to the migration of these intertidal environments, and therefore, the intertidal zone is becoming narrower as they are 'squeezed' between the rising sea and the sea walls. This adversely affects intertidal wildlife, such as wading birds, because the intertidal area, and therefore their feeding grounds, are diminishing. Many important wildlife sites may be lost by such coastal squeeze, but in eastern England some of the sea walls are being deliberately breached to allow the intertidal area to expand in a landwards direction.

on the increasing levels of CO_2 in the atmosphere from human activity, which is leading to an enhancement of the Greenhouse Effect, resulting in global warming. It must be pointed out that global warming may not be solely due to anthropogenic effects and that natural phenomena may be contributing, such as variations in solar radiation output.

Table 5.3 *IPCC (1996) estimated contributions to sea level rise (cm) during the period 1890–1990.*

	Low	Middle	High
Thermal expansion	2	4	7
Glaciers/ice caps	2	3.5	5
Greenland ice sheet	−4	0	4
Antarctic ice sheet	−14	0	14
Surface and ground water	−5	0.5	7
Total	−19	8	37
Observed	10	18	25

Regardless of the cause, global warming does appear to be happening and this will continue to affect global sea level (Gornitz 1995). Evidence from tide-gauge stations shows that eustatic sea level has risen by between 10 and 25 cm in the period 1890–1990 (Warrick *et al.* 1996). Five major climate-related factors have been recognised that could possibly explain this rise in global mean sea level. These factors are discussed in the following sections and in Table 5.3. The best estimate total (middle) amount of sea level rise resulting from the contribution of these five factors is 8 cm, which is below the observed amount and recognised as probably representing an underestimate.

Thermal expansion of the oceans

The volume of the oceans is strongly influenced by the density of sea water, which is inversely related to temperature. Therefore, as the oceans warm up due to global warming, the density of sea water decreases, leading to an increase in water volume, expansion of the oceans, and ultimately a rise in sea level. Sea level change resulting from density variations is called **steric sea level change**. Such steric changes may also be brought about by salinity variation, which although may be important locally, especially at the coast, the effect is relatively minor on the global scale. Estimating oceanic expansion can either be achieved through field observations and recording, or predicted using models. Thomson and Tabata (1987), for example, examined a steric height record over 27 years from the northeast Pacific and found that steric heights are increasing at approximately 0.9 mm y^{-1}. However, the accuracy of this, and other similar studies, is questionable because interannual variability produces considerable 'noise' in the record to be confident, and also the regional nature of the study makes inference on a global scale problematic. In order to fill information gaps and to overcome problems at the regional scale, the World Ocean Circulation Experiment (WOCE) has been designed in the long term to provide more accurate and widespread data. For the present though, in the absence of good observational data, models have been designed to predict steric changes. Wigley and Raper (1993) used such a model and suggest that for the period 1880–1990, the resultant range of sea level rise due to thermal expansion of the oceans is in the order of 3.1–5.7 cm. However, the encompassing figures cited by Warrick *et al.* (1996) for sea level rise for this period is 2–7 cm (Table 5.3).

The melting of glaciers and small ice caps

A number of studies have shown that the majority of non-polar valley glaciers have been retreating since the second half of the nineteenth century, some by several kilometres. This is most likely caused by the rising of altitudinally controlled climate zones found in mountainous regions, so that at the foot of mountain ranges, where many glaciers emerge, temperatures and melting rates are increasing. Meier (1984) estimated that during the period 1900–1961 glacier retreat contributed 2.8 cm, or 0.46±0.26 mm y^{-1}, to global sea level rise. This estimate has been superseded by Warrick *et al.* (1996) who suggest that between 2 to 5 cm has been contributed to sea level rise in the period 1890–1990 (Table 5.3).

The Greenland ice sheet

Estimating the contribution of the Greenland ice sheet to sea level rise over the past century is hampered by a lack of information regarding the dynamics of the ice sheet. It has been suggested that as global temperature increases, precipitation would increase over Greenland, leading to an increase in ice accumulation. Indeed, for every 1°C rise in global temperature, precipitation would increase by 4 per cent (Warrick and Oerlemans 1990). However, this accumulation is thought to affect the interior of Greenland only, as most of the outlet glaciers around the Greenland coast retreated strongly during the twentieth century. Therefore, it is not certain how the Greenland ice sheet has contributed to sea level, and Warrick *et al.* (1996) allow for both negative and positive contributions with an estimate of between −4 and 4 cm, a figure that is based on a sea level change rate of ±0.4 mm y^{-1} through 1890–1990 (Table 5.3). It seems probable, however, that its contribution is less than from thermal expansion and the melting of non-polar valley glaciers.

The Antarctic ice sheet

In a similar manner to the Greenland ice sheet, the Antarctic ice sheet is thought to accumulate ice through an increase in precipitation brought about by global warming. However, unlike Greenland, a number of studies have indicated that the Antarctic ice sheet has a positive balance, that is it accumulates more ice than it loses through melting. For example, Budd and Smith (1985) estimate that of the ice accumulated each year in Antarctica approximately 10 per cent is retained and not lost to melting, resulting in the growth of the ice sheet at a rate in the order of 209×10^{12} kg y^{-1}. This positive balance corresponds to a rate of eustatic change of approximately −0.6 mm y^{-1}, and therefore, contributing negatively to global sea level change. Other studies, however (e.g. Huybrechts 1990), suggest the contribution to sea level is positive by approximately the same amount. Whatever the contribution, it is finely balanced because the West Antarctic Ice Sheet (WAIS) is grounded below sea level, so that it may be very sensitive to sea level rise. Sea level rise may lift the WAIS off the sea floor by flotation. This would decrease the bottom stresses of the WAIS allowing it to flow faster and thin more rapidly. Iceberg calving would increase, which would result in a positive contribution to sea level rise. Warrick *et al.* (1996), with great uncertainty, estimate that the Antarctic ice sheet has contributed −14 to 14 cm to sea level change during the period 1890–1990 (Table 5.3).

Variations in surface and groundwater storage

Surface storage includes water stored in the soil, rivers, lakes and inland seas, an example being the construction of many dams and reservoirs since the 1930s, which have made a negative contribution to global sea level rise, perhaps by as much as −5.2 cm (Newman and Fairbridge 1986). Deforestation, however, often leads to water loss from soils through increased surface run-off and evaporation, which constitutes a positive contribution to sea level rise. Groundwater is water found in porous and permeable rock aquifers and has been extracted through pumping for domestic, industrial and agricultural purposes. Groundwater pumping amounted to the extraction of over 2,000 km^3 of water during the twentieth century, which is equivalent to a 0.55 cm rise in sea level (Warrick and Oerlemans 1990). Warrick *et al.* (1996), again with uncertainty, estimate a contribution to sea level of between −5 and 7 cm during 1890–1990 (Table 5.3). More recently, Gornitz *et al.* (1997) support a negative contribution to sea level rise from these sources.

Future sea level rise and its impacts

The results of a number of model experiments reported by Warrick *et al.* (1996) indicate that a sea level rise of 49 cm by 2100 is possible, but with a range of uncertainty of 20–86 cm, and represents a rate of sea level rise that is significantly faster than that experienced over the majority of the twentieth century. These projections of future sea level rise are presented under a number of scenarios that use different levels of greenhouse gas emissions. The scenarios are labelled IS92a–f, and IS92a is equivalent to the 'business-as-usual' scenario of Houghton *et al.* (1990), which is a projection using the present greenhouse gas emission levels. The other scenarios are for a number of lower or higher emission cases. For each scenario, a high, low and best-estimate is calculated, each associated with a global atmospheric temperature rise of 4.5°C, 1.5°C and 2.5°C respectively.

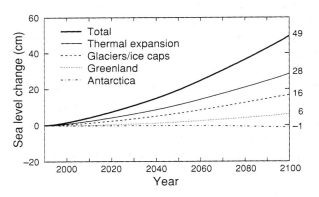

Figure 5.8 *Future sea level rise to 2100, indicating total and constituent contributors.*
Source: reproduced with permission from the Intergovernmental Panel on Climate Change from Warrick *et al.* 1996 (Figure 7.8, p. 384)

The various contributors to sea level change in the period 1990–2100 have been assessed individually (Figure 5.8). A thermal expansion estimate of 28 cm has been calculated using the computer model of Wigley and Raper (1993). A 16 cm contribution from mountain glaciers and small ice caps was calculated in relation to changes in glacier volume in response to global warming; the warmer the climate, the greater degree of melting, leading to a decrease in glacier volume. The Greenland ice sheet is estimated to contribute 6 cm, whilst the Antarctic ice sheet is likely to contribute negatively by −1 cm. In summary, under the IS92a scenario, the best estimate for the year 2100 is that global sea level would have risen by 49 cm, although because of the range of uncertainty it could be as little as 20 cm or as much as 86 cm.

The projections also show that throughout the twenty-first century, even under the scenarios of lower greenhouse gas emissions, sea level continues to rise. This suggests that even if greenhouse gas emissions were cut completely now, sea level would continue to rise. This is mainly because of a lag effect caused by the ability of the oceans to store heat, in what is called its thermal inertia. This means that society is committed to sea level rise. However, global sea level rise will have different impacts on different coastlines, mainly due to isostatic movements, but there is considerable concern for the majority of the world's coasts (see Box 5.4). Concern is not centred on the day-to-day increase in tidal range, which can in most cases be contained, but in the increased ability of extreme events, such as storm surges and very high tides, to flood coastal lowlands and cause extensive coastal erosion. Deep storm depressions are predicted to become more frequent under global warming, so that increased sea levels, coupled with surge conditions and high precipitation in coastal lowland river catchments, are inevitably going to mean higher frequency and higher magnitude flood events.

Box 5.4

Socio-economic impacts of future sea level rise

Bijlsma (1996) on behalf of the Intergovernmental Panel on Climate Change outlines the socio-economic implications of predicted sea level rise due to an enhanced Greenhouse Effect. These include:

- General negative impacts on various coastal amenities and activities, such as tourism, freshwater supply and quality, fisheries and aquaculture, coastal agriculture, human settlements, financial services and human health.
- Protecting low-lying islands, such as the Maldives and Marshall Islands and countries with extensive low-lying deltaic plains, such as Bangladesh, Nigeria, Egypt and China, will probably be extremely costly.
- The numbers of people globally who currently suffer from coastal flooding (40 million) could double or triple by 2100.
- Attempting to adapt to sea level rise may mean that some societies will face difficult decisions regarding which coasts to protect and which to abandon, and trade-offs between environmental, economic, social and cultural values will have to be made.

Furthermore, Houghton (1994) indicates that for the United States alone, the financial cost of coastal protection and land loss will amount to US$7 thousand million, which doesn't include the cost from seawater contamination of water supplies, agricultural losses, and health problems that may be caused by sea level rise. This figure is approximately 1 per cent of the US Gross Domestic Product (GDP), and a similar percentage is predicted for other developed countries, whilst for some developing countries costs could be as much as 6 per cent of GDP.

Managing global sea level rise

Carter (1988) realised that predicted sea level rise into the twenty-first century will lead to coastal erosion, redistribution of sediments, wetland submergence, floodplain waterlogging and salt contamination of coastal aquifers. Disruption will be caused to residential, industrial and commercial activities, transport routes will be severed, and agricultural land ruined. Thus, with global sea level rise there is going to be an increasing demand for management solutions to protect thousands of kilometres of coast, and addressing these consequences will require a major redistribution of money at both the national and international levels. Good coastal protection schemes, such as the Thames Barrier in London, are not viable on a worldwide basis, and governments need to accept the loss of coastal land. The early anticipation of sea level rise is of great value because it allows planning to be introduced gradually.

Various national organisations have instigated policies and plans to address the sea level rise management issue. Bray *et al.* (1997) provide advice to coastal management decision-makers and suggest a number of planning options for the coast of southern England:

- *Retreat* – this has a number of meanings, including the practice of **managed retreat**. It may involve the relocation of coastal communities or industry, with a prohibition on further development. This would require a strong governmental role with supportive legislation, and with the associated expense of relocation costs and compensation. Coastal processes would operate naturally in response to sea level rise, but realistically only undeveloped, low-density or precarious land would be left to retreat in this way. Alternatively, retreat may involve 'conditional development' whereby the coast continues to be used, but with conditions in leases set by government that will eventually enable the sea to reclaim the area. The final retreat option inolves the surrendering of governmental responsibility who, after stating the risks, abandon the coast to private market forces.
- *Accommodation* – this essentially means learning to live with sea level rise and coastal inundation. This may be through an adaptation to the effects of sea level rise, such as elevating buildings, enhancing storm and flood warning systems and modifying drainage. For communities already used to coastal flooding, or where retreat or protection are not socio-economically possible, this may be an ideal option, but it does require high levels of organisation and planning and community participation, all of which can be quite costly to implement. It may be possible to change activities, such as changing agricultural practices, to suit the new environment, or simply to accept the risks of inundation and accept increasing insurance premiums. This final option is only really viable in urban areas perceived to be at low risk from coastal inundation.
- *Protection* – this involves the physical protection of the coast, such as the construction of 'hard' structures (e.g. sea walls, groynes, etc.), for developed

stretches of the coast that are particularly vulnerable to erosion or inundation. A disadvantage of 'hard' protection is that it is fixed, and if the shoreline does change, a new line of defences would have to be built, representing a duplication of the original considerable cost. The alternative method of protection is known as the 'soft' option, which involves artificial beach nourishment for example, which is commonly employed, ironically, to afford a measure of protection for 'hard' structures. These 'soft' options often mimic natural processes, but their maintenance can be costly, especially if sediment transport removes introduced sand from the beach system.

The option adopted by managers and planners of any given coast will be influenced by human factors, such as population, settlement, business, and industry (i.e. socio-economic), more than physical factors. Figure 5.9 illustrates schematically the relationship between the value of a coast to society (which may not always refer to its economic value) and the cost of protecting that coastline. It indicates that where the value of the coastal land is less than the cost of protecting it, then communities will either have to accommodate the effects of sea level rise, or accept retreat.

It has also been suggested that it may be possible to offset global sea level rise by transferring sea water on to land. Candidates for this flooding include the Dead Sea region in the Middle East, which lies well below present sea level, with the advantage that hydro-electric power could be generated in the water transfer process from sea to land (Carter 1988). However, schemes like this are perhaps not likely to reach fruition due to local socio-economic and political conditions.

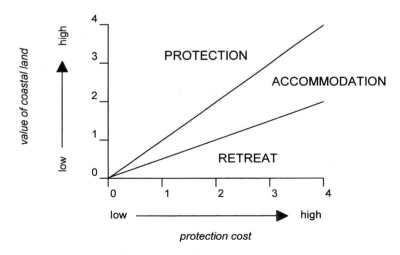

Figure 5.9 *The relationship between the value of coastal land and the cost of protecting that coastline. Note: The numbers on the axes are intended to indicate comparable value and cost, although the value of coastal land may not always refer to its economic value.*

Summary

- Sea level is not constant in either space or time.

- Relative sea level represents the balance between eustasy and isostasy, both of which are strongly influenced by the growth and decay of continental ice sheets through the Quaternary.

- Sea level curves depicting sea level change are best constructed using sea level index points for which age, altitude, tendency of sea level and indicative meaning and range are known.

- The response of coastlines to sea level change varies, but landward migration is likely for depositional coasts where the sediment is mobile and/or in abundant supply. Immobile coasts become submerged and drown.

- Sea level appears to have risen throughout the twentieth century and is predicted to continue during the twenty-first century as a consequence of human activity and the enhancement of the Greenhouse Effect.

- As a consequence of predicted sea level rise, coastal societies must either: (a) learn to accept the loss of coastal land; (b) accommodate sea level rise; or (c) protect coastlines.

Discussion questions

1 Discuss the method of construction, and evaluate the accuracy of Holocene sea level curves.

2 A number of climate-related factors were recognised by the Intergovernmental Panel on Climate Change (1995) that could possibly explain a rise in global mean sea level on the 100-year scale. Examine the sensitivity of these factors to changes in climate and outline how they have stimulated sea level change over the last one hundred years.

3 Examine the variety of options open to coastal populations in response to future sea level rise.

Further reading

See also

The tectonic classification of coasts, Chapter 1.
Coastal systems and sea level examples, Chapters 2 and 4.
Coastal management issues, section Chapter 6.

General further reading

Sea-Level Changes: The Last 20,000 Years. P. A. Pirazzoli. 1996. Wiley, Chichester, 211pp. A well-written introduction to sea level changes on a geological timescale.

Ice Age Earth: Late Quaternary Geology and Climate. A. G. Dawson. 1992. Routledge, 293pp. An extensive review of climate and environmental change from the last glacial onwards.

Sea-level Research: A Manual for the Collection and Evaluation of Data. O. van de Plassche (ed.). 1986. Geo Books, Norwich, 618pp. The standard reference work for the collection and interpretation of sea level data.

Submerging Coasts: The Effects of a Rising Sea Level on Coastal Environments. E. C. F. Bird. 1993. Wiley, Chichester, 184pp. A useful review of the response of coastlines to rising sea levels.

Climate and Sea-level Change: Observations, Projections, and Implications. R. A. Warrick, E. M. Barron and T. M. L. Wigley (eds.). 1993. Cambridge University Press, Cambridge, 424pp. A collection of papers examining the issue of global climate change and associated sea level rise.

Impacts of Sea-level Rise on European Coastal Lowlands. S. Jelgersma and M. J. Tooley (eds). 1992. Blackwell, Oxford, 267pp. An example of how sea level rise impacts on a regional scale.

6 Coastal management issues

Coasts are some of the most populated areas in the world, and the interaction between humans and the environment often throws the natural coastal system out of equilibrium. Management is required to mitigate against these human impacts on the environment and upon society. This chapter covers:

- **sustainable coastal zone management**
- **a range of coastal management issues, such as population growth, coastal uses, coastal hazards and administrative issues**
- **a number of approaches to intergrated coastal zone management**
- **various examples of human impacts upon the coastal environment**

Introduction

Hopefully this book, if nothing else, has imparted an appreciation of the sensitivity of coastal systems, and that intervening in one part of a system will have an effect on the whole system. Human activity at the coast invariably intervenes in coastal systems, often with detrimental impacts. In order to minimise these impacts, which affect both the natural environment and the intervening human population, coasts must be managed. Management is also concerned, however, with non-anthropogenic processes that affect coasts, but impede on human activity, such as wave attack and coastal erosion. Therefore, the aims of coastal management are to facilitate the human use of the coastal zone, but minimising the impacts of such human use, and to protect human interests at the coast from natural and human-related processes.

Sustainability is now a key word in environmental management, so that all human use of the coast should meet the needs of the present population without jeopardising the opportunities of future generations, so that our descendants will be able to use the coast in much the same way as we do today. Sustainable coastal management, therefore, means that all human activities should be non-destructive and that the resources we exploit are renewable. Against such a benchmark, it is clear that many practices are not sustainable and that our descendants will not inherit the opportunities that the present generation enjoys. However, we are at a critical time for the management of the global environment, and if sustainable management practices are widely adopted now, the opportunities our children inherit will be maximised.

Coastal management issues

The issues that coastal management is expected to address are very diverse and mostly local or regional in extent. However, some coastal management issues do affect the global coastline, such as the threat of future sea level rise as a result of global warming discussed in Chapter 5. Nevertheless, all these issues must be appreciated by the coastal manager so that holistic responses can be made. This, of course, is in addition to an intimate understanding of the operation of coastal systems, which has been the main focus of this book. Kay and Alder (1999) discuss at length the variety of coastal management issues and employ many excellent case studies. The following sections examine these issues.

The growth of coastal populations

During the time it has taken to write this book the global population passed the 6 billion mark, and over 50 per cent of these people live in the coastal zone – that is equal to the entire global population of the 1950s. The future predictions are just as alarming in that by 2025 more people will live in the coastal zone than were alive in the 1990s. This incredible pressure placed upon the coast by an ever-increasing global population is the principal reason why coasts are under threat, and why they need the level of management and protection that they do.

This increase in coastal populations is not restricted to developing countries, with some 50 per cent of people in developed countries living less than 60 km from the coast. The reasons for this population concentration at the coast are complex, but include:

- The general global population increase, with indigenous coastal populations growing.
- Rural to urban migration. Rural depopulation is currently of global concern, and because many of the world's major cities are sited on the coast, they are attracting people from displaced rural populations.
- Inland to coastal migration. This concerns the migration of people from inland areas specifically and intentionally to the coast for economic, social and recreational reasons. In developed countries this is often brought about by people retiring to the coast.

The human geography of coastal cities is of significance in understanding coastal population growth, in that most major coastal cities have developed as a major port. This allows the import and export transport of material and goods, and encourages industry and business to develop at or near the port site. Such ventures stimulate employment and investment opportunities, both of which attract people to come and work and live in the coastal city. Under these circumstances, expanding coastal populations put immense pressure on local planning authorities to provide suitable land for urban and further industrial development, so consuming more coastal land. This can form the basis of a cycle which leads to the continous increase in city size. Indeed, some neighbouring coastal cities have been known to merge in this way, such as Tokyo and Osaka in Japan (Kay and Alder 1999).

Box 6.1

The impact of development in the Florida Keys (USA)

The Florida Keys are an extensive coastal wetland in the southern USA. The coast comprises barrier islands, mangroves, sea grass beds and coral reefs. The area is a popular retirement region, due to its generally good weather, easy-going lifestyle and diverse recreational activities. Tourism is also a growing industry here. Development of the Keys has expanded greatly in recent years so that some mainland and barrier island shorelines are completely given over to residential development, and little of the natural shore remains. Deford (1999) outlines some of the impacts that this continued residential development and recreational activity has had on the mangrove, sea grass and coral reef environments.

- *Mangroves* – these are extremely important to the Florida Keys marine environment, as they act as nurseries for many species of fish and spiny lobster, their roots filter pollutants out of the sea water and they serve to buffer the coast against hurricanes. Although being protected by law, mangroves are still being deforested by landowners and developers in order to improve waterfront views and to create docks for pleasure craft.
- *Sea grass beds* – these occur in shallow water seaward of the shoreline and are important in trapping sediment which may otherwise be transported further seaward where it would be harmful to coral. They also absorb excess nutrients in the water from poorly treated sewage and storm water run-off from urban areas. However, if nutrient levels are too high algal blooms occur that block off sunlight reaching the sea grass, killing them. Pleasure boat propellers create grooves in the sediment which are difficult for sea grass to recolonise, and similar effects are produced by anchors and lobster pots dragged across the beds. Dredging of channels undermines sea grass roots, so promoting localised erosion and releasing sediment which requires further dredging.
- *Coral reefs* – these are important in dampening the effect of waves at the shoreline, protecting the landward sea grass beds and mangroves. And yet they suffer physical damage from anchors, grounded vessels and from souvenir collecting. Pollution affects the reefs through algal blooms limiting light and reducing the activity of the photosynthesising coral symbionts, thus increasing disease and coral mortality.

The Florida Keys are managed by the Florida Coastal Management Program (FCMP) and operates under the Florida Coastal Management Act of 1978. The FCMP administers the Florida Keys National Marine Sanctuary, and contributes towards schemes set up to help mitigate against the threat of hurricanes, and to implement hurricane evacuation plans (Kay and Alder 1999).

The use of the coast

The use of the coastal zone is probably as old as our species and comprises extremely diverse practices. Besides the use of coastal land for industrial and residential development (discussed above), coastal use may be split into four categories: resource exploitation, infrastructure, tourism and recreation, and conservation (Kay and Alder 1999).

Resource exploitation

This includes the use of the coast's physical and biological resources. Although ideally such exploitation should be done on a sustainable basis, this has not generally been the case. Coastal fish stocks have declined with over-fishing, so that many species are depleted or indeed fully fished. The relatively recent rise of aquaculture on a large scale, for fish, prawns, seaweed and sea cucumbers, has resulted in the loss of natural coastal wetland for pond creation, and major pollution problems from the high levels of nutrient output (unconsumed food, faecal pellets, etc.) often leads to coastal eutrophication, especially where the aquaculture is based on caged stock in the coastal sea. Such aquaculture may also lead to the increase of disease in marine organisms, and the inevitable accidental release of exotic species may displace native populations.

Mangrove deforestation is also a major problem for tropical coastlines. Traditionally mangroves have been used for firewood and construction material, but the present high demand placed upon these resources by growing coastal populations is becoming unsustainable. Also, major tracts of mangroves are being converted to aquaculture. Uniquely, 50 per cent of the mangrove forests of the Mekong Delta were destroyed during the Vietnam War by the US Army use of chemical defoliants and napalm (Viles and Spencer 1995). The impact of this activity is to reduce coastal biodiversity and sedimentation, and to leave the coastline more vulnerable to cyclone events and flooding.

The exploitation of oil and gas, and of other materials that are mined in the coastal zone is entirely unsustainable, as it involves the permanent removal of resources. Deltas are often rich in hydrocarbon reserves and many major oil and gas fields are situated on or near deltaic systems, such as the Mississippi Delta in the southern USA. The construction of oil rigs and pipelines often is large scale and involves degradation of the coastal zone. Mining of coastal sediment through quarrying and dredging is practised for supplying aggregate for the building industry (Plate 6.1), deepening shipping channels and for supplying sand for beach nourishment schemes.

Plate 6.1 *Quarrying of dune sand and gravel in St Ives Bay, Cornwall (UK) in 1996.*

Box 6.2

The impacts of sediment extraction

Sediment extraction for aggregates may destabilise coastal systems and lead to long-term shoreline retreat. For example, in the Baie d'Audierne (Brittany, France) the shoreline has retreated over 50m since *c.* 1945. Plate 6.2 clearly shows the position of a German Second World War gun emplacement presently lying in the middle of Plage de Tronoan in the southern part of the bay. Guilcher *et al.* (1992) describe how these military installations were originally located along a high gravel barrier which controlled the recession of the coastline. The gravel barrier suffered from sediment extraction both during the war and afterwards which left it susceptible to breaching by waves. Sand was washed through breaches during storms to form deltas over 200 m wide on the landward side of the barrier. Onshore winds transported this sand further inland to form dunes which rapidly became stabilised by vegetation. In this case, there has not only been extreme coastline retreat, but also a change in the coastal environment. There is now no sign of the former gravel barrier at Plage de Tronoan and the narrow belt of low dunes present today are showing signs of continued retreat.

Continued

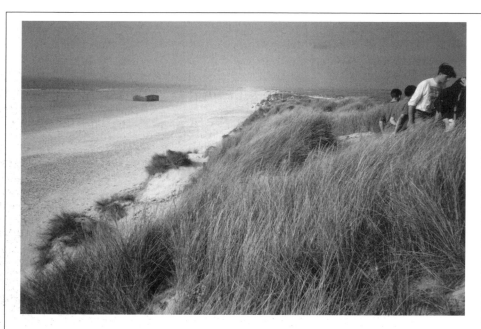

Plate 6.2 *Plage de Tronoan in the Baie d'Audierne (Brittany, France) in 1994. The blockhouse on the beach was built by the German Army during the Second World War on top of a gravel barrier. Gravel extraction during the Second World War and after led to the destruction of the barrier, retreat of the shoreline and construction of a dune system.*

Infrastructure uses

This refers to the building and expansion of ports, with associated shipping, industrial and urban development, including road and bridge building (Plate 6.3). With the advent of very large ocean-going vessels, ports have had to respond by deepening shipping channels and providing deep-water docking facilities. The ability to transport increased cargo loads has stimulated industrial expansion and the development of major urban centres at the coast. This has led to the building of roads and bridges to facilitate the distribution of imported and manufactured goods, and to allow exports to reach the port. The demand for coastal land for development in these situations is great, and has prompted coastal wetland reclamation and even the construction of artificial land. Pollution is a major problem in ports, with some of the worst pollution events involving oil spills from foundered tankers. Unfortunately, there are many examples of this, such as the *Sea Empress* disaster in the port of Milford Haven (UK) in 1996.

Plate 6.3 *Infrastructure uses of the coast are well illustrated by the view of the port of Penzance in Cornwall (UK). The area is well built up and very little undeveloped coast remains in the immediate vicinity.*

Box 6.3

The problem of coastal litter

It is almost impossible to visit a beach now without seeing some litter (Plate 6.4). Many coastlines around the world are plagued by litter, and this is posing great problems to health, scenery, wildlife and the tourist industry. Many municipal and voluntary schemes try to clear beaches of litter, but it appears to be a losing battle for fresh litter usually returns upon each high tide. It is now widely appreciated that in order to tackle the problem of beach litter, the origin must be identified, so that it may be stopped at the source. There are three potential sources of this litter:

1 *Marine sources* include rubbish thrown overboard by boats and ships. This is often the greatest source of foreign litter found on beaches.
2 *Beach sources* refers to direct littering on to a beach, most often by beach-users, mainly tourists.
3 *Riverine sources* comprise litter washed into the sea from rivers, and may be from sewage outlets and fly-tipping (illegal disposal of waste) on river banks.

Williams and Simmons (1997) have analysed litter on beaches in the Bristol Channel (UK) and found some very alarming amounts of litter, the majority of which they

Continued

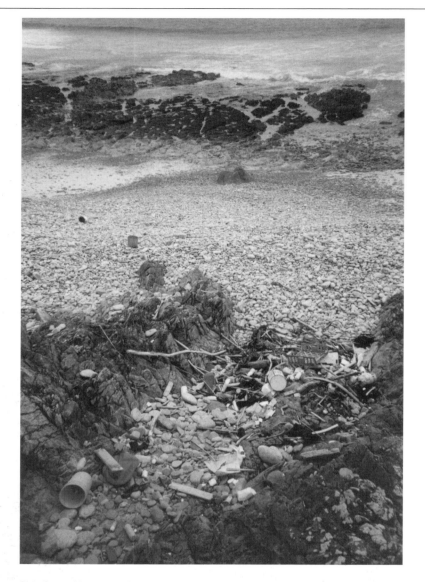

Plate 6.4 *Strandline deposits of litter on Plage de Mezpeurleuch (Brittany, France) – an all too familiar sight on beaches today.*

attribute to discharge into the sea from rivers, such as the River Taff. For example, Merthyr Mawr beach was found to contain 550 plastic bottles, 75 sanitary items and 210 cans per kilometre of strandline.

Tourism and recreation

Tourism is perhaps one industry which could, if managed properly, be compatible with environmental sustainability. However, it is often poorly managed, both in developed and developing countries, and leads to the degradation of coastal sites, reducing their suitability as tourist destinations and resulting in economic decline. Tourist facilities, such as caravan parks (Plate 6.5), camping sites, resorts, golf courses and marinas, are often installed without full appreciation of coastal systems. Common impacts of tourist and recreation activity include:

● Coastal sand dune degradation, where caravan and camp sites are built behind dunes, pathways cross the dunes that incise with continued use, leading to the undermining of dune vegetation and blow-out formation. Horse riding and on-dune driving can also have the same effect. This may ultimately result in the fragmentation of the dune system, leaving the land behind the dune vulnerable to inundation during storms.
● Heavily used coastal cliff-top footpaths, if not properly managed, can be eroded causing slope instability and soil degradation (Plate 6.6).
● Artificial marinas can provide new quiet water conditions that act as a sink for sediment deposition, therefore removing sediment from longshore drift systems, which may leave downdrift sections of coastline vulnerable.

Plate 6.5 *Caravans and motor-homes parked on a salt marsh at Mont St Michel (Normandy, France), the weight of which is affecting marsh cliff stability.*

Plate 6.6 *Pointe du Raz in Brittany is the westernmost tip of mainland France and suffers intense tourist pressure, culminating in the summer with 2,000 visitors per hour which has had serious environmental impacts, especially soil erosion. Measures to combat these impacts have taken the form of building the visitor centre over 1 km away from the site,*

- Water craft can damage the sea bed, such as outboard motor propellers that damage sea grass beds and the deployment of anchors that damage coral.
- The outflow of sewage into the sea, which in many developing countries is poorly treated, if at all. This pollution causes environmental damage through eutrophication and poses severe health risks for the tourist population.

These are just some potential impacts of tourist activities on the coast, and does not include the construction of tourist accommodation which can consume large tracts of coastal land.

Conservation uses

There are many nature reserves and wildlife parks situated at the coast. However, Agenda 21 of the Rio Earth Summit of 1992 requires that further measures be taken to safeguard global biodiversity. Kay and Alder (1999) suggest that Agenda 21 requirements from the coastal zone will be met through the planning of new conservation areas. Indeed, new coastal developments must often address the conservation issue, and provide 'buffer' zones which act as transitional spaces from developed to natural spaces, that allow physical processes to operate. The creation of new reserves may often simply involve the protection of natural coastal environments. There are cases, however, where the conservation interest in a coastline is at odds with the operation of the natural system. This is particularly so for waterfowl reserves in coastal wetlands, where grazing, nesting and feeding habitats are needed, so that some areas should be permanent water, whilst others semi-permanent marshy grassland. In order to obtain this landscape diversity in a coastal wetland, the water table must be controlled, embankments built around some areas for pasture and drainage ditches dug.

Coastal hazards

Coastal hazards are important management issues as it is this group of issues that cause obvious environmental and social disasters. These include short-term events, such as the landfall of cyclones, hurricanes and typhoons, and the impact of tsunami, and longer-term issues such as wave activity and the predicted gradual rise in future sea level that may be associated with climate change (see Table 6.1 for a summary of coastal protection measures). Managing the coast to combat these

providing designated paths and restricting access to certain areas. Photo (a) was taken in March 1997 shortly after the start of the rehabilitation programme and (b) in March 1999 indicating the level of vegetation regrowth that has occurred.

Box 6.4

Coastal eutrophication

Eutrophication occurs when excess nutrients enter the aquatic environment and cause algae to bloom, which in turn depletes oxygen in the water. The types of nutrients that may trigger eutrophication include nitrates and phosphorus, and these may be sourced from agricultural fertilisers, poorly treated or untreated sewage and animal wastes. In the coastal waters of Brittany (France) the problem is quite severe, it being an intense agricultural region. Blooms of *Ulva* algae occur each year in sheltered embayments, such as the Baie de Douarnenez, and become deposited on beaches (Plate 6.7), the sight and smell of which seriously detract from the tourist value of the locations. These deposits are collected to clean the beaches, and the algae is re-used as fertiliser. In 1992, 82,000 m^3 were collected from Breton beaches (Fera 1993). Furthermore, coastal sedimentation is important as a sink for pollutants, especially mining wastes, radionuclides and trace metal contaminants. However, these sinks are often only temporary and pollutants can be disturbed and re-released by dredging activity and boat traffic, as well as by natural processes of erosion. In a number of cases, uncontrolled discharge of pollutants has led to the complete elimination of bottom-living organisms.

Plate 6.7 *Vast accumulation of algae on Fautec Beach (Brittany, France). An unsightly consequence of coastal eutrophication.*

issues often requires high levels of finance and long-range management plans. For example, combating flooding associated with storm surges has involved the construction of costly barrages across estuaries, such as the Thames Barrier in

Table 6.1 A summary of engineered coastal protection measures (see Carter 1988, for full discussion).

management issue	engineered solution(s)	types	description	problems	illustration
cliff erosion	seawalls	vertical wall	a wall constructed out of rock blocks, or bulkheads of wood or steel, or simply semi-vertical mounds of rubble in front of a cliff	rock walls are highly reflective, bulkheads less so. Loose rubble however, absorbs wave energy	
		curved wall	a concrete constructed concave wall	quite reflective, but the concave structure introduces a dissipative element	
		stepped	a rectilinear stepped hard structure, as gently sloping as possible, often with a curved wave-return wall at the top	the scarps of the steps are reflective, but overall, the structure is quite dissipative	
		revetment	a sloping rectilinear armoured structure constructed with less reflective material, such as interlocking blocks (tetrapods), rock filled gabions, and asphalt	the slope and loose material ensure maximum dissipation of wave energy	
coastal inundation	seawalls	earth banks	a free-standing bank of earth and loose material, often at the landward edge of coastal wetlands	may be susceptible to erosion, and overtopped during extreme high water events	
	tidal barriers		barriers built across estuaries with sluice gates that may be closed when threatened by storm surge	extremely costly, and relies on reliable storm surge warning system (e.g. Thames Barrier)	
beach stabilisation	groynes		shore-normal walls of mainly wood, built across beaches to trap drifting sediment	starve downdrift beaches of sediment	
	beach nourishment		adding sediment to a beach to maintain beach levels and dimensions	sediment is often rapidly removed through erosion and needs regular replenishing; often sourced by dredging coastal waters	
offshore protection	breakwaters		structures situated offshore that intercept waves before they reach the shore. Constructed with concrete and/or rubble	very costly and often suffer damage during storms	
tidal inlet management	jetties		walls built to line the banks of tidal inlets or river outlets in order to stabilise the waterway for navigation	the jetties protrude into the sea and promote sediment deposition on the updrift side, but also sediment starvation and erosion on the downdrift side	

London (UK). In Japan, where tsunamis occur relatively frequently, many coastal towns have built high walls and gates around key areas that will act as a retreat if a tsunami approaches. In a similar manner, elevated platforms are constructed along some vulnerable low-lying coasts so that people can escape above the flood waters. The management responses to predicted sea level rise have already been discussed on pp. 159–160. In addition to these extreme events, above average wave activity can accelerate coastal erosion and cause cliff retreat, destroying cliff-top buildings.

Administrative issues

The diversity of problems facing the coast and the varying temporal and spatial scales at which they operate, inevitably bring in many different organisations with an interest in the management of a coastline. This may typically include local, regional, national and international administrative authorities, such as councils and governments, environmental and conservation organisations and other groups which may represent residents, tourist and industrial interests. It is not unusual along any coastline for a mosaic of different coastal managers to exist. There is a strong argument, however, that for effective management of a coast, an integrated approach should be adopted, either by all small concerns working closely together to decide widely applicable management plans or by amalgamation so that one body has overall responsibility for management planning.

Cicin-Sain (1993) outlines a continuum of coastal management integration (Kay and Alder 1999). Five stages along this continuum have been described:

1 *Fragmented stage* refers to many little organisations operating independently with little communication between them.
2 *Communication stage* where there is regular but occasional communication between different organisations.
3 *Coordination stage* involves different organisations in close communication and synchronising their work.
4 *Harmonisation stage* where different groups synchronise their work along universally agreed policy guidelines.
5 *Integration stage* refers to complete synchronisation and formal mechanisms along which work must be carried out, and therefore, independence of individual organisations is lost and fully integrated coastal management is achieved.

As one moves towards integration along this continuum there is usually a high degree of legislation that allocates management responsibility to one particular organisation, which may or may not be governmental.

Approaches to coastal zone management

From the previous section it is clear that management of the coastal zone operates at many different levels from the local to national. It is often the financial resources, and legal authority and responsibility at these different levels that determine the form of management employed. With such a diversity of management approaches practised, it is impossible here to detail all, and therefore, only a small number of approaches are reviewed to serve as examples.

SCOPAC – coastal zone management in southern England

In Britain, the Coastal Protection Act 1949 states that the protection of the coastline from erosion is the responsibility of the local authority, and protection from flooding is within the remit of the Environment Agency. Therefore, coastal management in Britain has been very fragmented. In 1986, the local authorities of the southern England coast realised the need for closer cooperation between neighbouring authorities. This realisation came in the light of increasing shoreline erosion problems and the *ad hoc* nature of local management solutions, which often had adverse effects along an adjacent authority's coast. As a result, the Standing Conference on Problems Associated with the Coastline (SCOPAC) was created, and included the coast between Dorset in the west and West Sussex in the east. Membership includes officers and elected members of district and county councils, the Environment Agency, harbour authorities and English Nature, the national nature conservation body. The full conference meets three times a year, local authority officers at least another three times, and specialist sub-group meetings also occur (Hooke and Bray 1995).

SCOPAC has a number of aims and objectives:

- to ensure a fully coordinated approach to all coastal engineering works and related matters between neighbouring authorities . . .;
- to eliminate the risk of coastal engineering work carried out by one authority adversely affecting the coastline of a neighbouring authority;
- to exchange information on the success or failure of specific types of coastal engineering projects;
- to establish close liason with Government and other bodies concerned with coastal engineering projects;
- to identify aspects of overall coastal management where further research work is required and to promote such research.

(Hooke and Bray 1995: 359)

In order to facilitate the last of these aims, each local authority makes financial contributions to SCOPAC for the funding of research. Some of the funds are also used to influence coastal management policy decisions at the Governmental level, and has even included the organisation of field trips to which Members of Parliament have been invited to observe coastal problems at first hand. SCOPAC has been successful in fulfilling most of its aims, and this has prompted other authorities around the British coastline to create local groups (Figure 6.1). The setting up of these coastal management groups has been done, as far as possible, according to the distribution of natural coastal cells, that is stretches of coastline which behave as essentially self-contained coastal sub-systems, with identifiable

Figure 6.1 *Coastal management groups of England and Wales.*

Source: reproduced with permission of the Royal Geographical Society from Hooke and Bray (1995) (Figure 1, p. 362)

inputs, transfers, stores and outputs of sediment (Hooke and Bray 1995). By delineating group boundaries coincident with natural systems boundaries, it is hoped that more holistic and fully integrated management may be possible. This approach to integrated coastal zone management has great potential for developing sustainable management plans.

Heritage Coasts of England and Wales

Management schemes such as SCOPAC address obvious needs for coasts that are under population, development, tourist and industrial pressure. However, unspoilt and scenic coastlines require management as well, and this has led in England and Wales to the setting up of Heritage Coasts which aim 'to conserve the quality of the scenery [and] to foster leisure activities based on the natural scenery rather than man-made features' (Williams 1992: 153). The Heritage Coasts scheme came about in 1973/1974 following reports by the National Parks Commission (NPC), and is based on voluntary agreements between local councils and landowners on how best to manage a given coastline. It is of very low budget, with 50 per cent of the financial contributions made by local councils, and a further 50 per cent by the NPC (and its successors the Countryside Commission, the Countryside Agency (England) and the Countryside Council for Wales). The money is used to employ a Heritage Coast Officer and rangers, and some limited funds for building and protection work. At present, 1041 km of coastline is designated Heritage Coast status in England, which is 32 per cent of the entire English coastline (Table 6.2) and some 500 km in Wales. Many of them have been deemed successful in their aims (Williams 1992) which demonstrates that non-statutory management schemes can work.

Coastal zoning – the Great Barrier Reef Marine Park

The zoning of the coast is a relatively simple and yet effective way of managing and separating incompatible uses. Activities in zones may be allowed, allowed with permission (i.e. to licence holders) or forbidden, and can be applied in economic, development, tourist or conservation situations. The Great Barrier Reef Marine Park (GBRMP) has adopted such a zonation strategy. Because of its size the park has been divided up into sections, so that the individual sections are more easily managed and human impact reduced. Examples of the kinds of zones defined for the Cairns section of the GBRMP include zones for general use, habitat protection, conservation parks, buffers, national parks and preservation (Figure 6.2; Kay and Alder 1999). Although it is agreed that such zoning is an excellent mechanism for managing the park on the broad scale, it requires additions at the local level. For example, in the Cairns section, the Green Island Management Plan was brought into

Table 6.2 *Heritage coasts of England.*

Coast	Defined (year)	Length (km)	County/region
Dover to Folkstone	1975	7	Kent
East Devon	1984	27	Devon
Exmoor	1991	45	Devon/Somerset
Flamborough Headland	1989	19	Yorkshire
Godrevy–Portreath	1986	9	Cornwall
Gribben Head–Polperro	1986	24	Cornwall
Hamstead	1988	11	Isle of Wight
Hartland	1986	11	Cornwall
Hartland	1990	37	Devon
Isles of Scilly	1974	64	Isles of Scilly
Lundy	1990	14	Devon
North Devon	1992	32	Devon
North Norfolk	1975	64	Norfolk
North Northumberland	1992	110	Northumberland
North Yorkshire and Cleveland	1981	57	North Yorkshire/Cleveland
Pentire Point–Widemouth	1986	52	Cornwall
Penwith	1986	54	Cornwall
Purbeck	1981	50	Dorset
Rame Head	1986	8	Cornwall
St Agnes	1986	11	Cornwall
St Bees Head	1992	6	Cumbria
South Devon	1986	75	Devon
South Foreland	1975	7	Kent
Spurn	1988	18	Yorkshire
Suffolk	1979	57	Suffolk
Sussex	1973	13	Sussex
Tennyson	1988	34	Isle of Wight
The Lizard	1986	27	Cornwall
The Roseland	1986	53	Cornwall
Trevose Head	1986	4	Cornwall
West Dorset	1984	41	Dorset
Total		1,041	

Source: reproduced with permission from the Countryside Agency's website
(http://www.countryside.gov.uk/what/hcoast/heri_tbl.htm)

effect in 1993 to formalise the use of the resources of Green Island, so as to permit recreational activity, but also to protect its natural assets (Figure 6.3; Plate 6.8).

Source: GBRMPA

Zoning Classifications
□ General Use Zone
▨ Habitat Protection Zone
▨ Buffer Zone
■ National Park Zone
▤ No Structure Subzone

N
0   5 km

Figure 6.2 *Coastal zoning approach to coastal management in the Cairns section of the Great Barrier Reef Marine Park (Australia).*
Source: Kay and Alder (1999) (Box 4.16, p. 190)

Box 6.5

Remote sensing and geographic information systems – high-tech tools for the coastal manager

Managing the coast involves monitoring changes and the handling of much information. These tasks have been aided in recent years by the expansion of remote sensing techniques and geographic information systems (GIS). Remote sensing involves making observations some distance from the subject and, as far as the coastline is concerned, includes aerial photography and satellite imagery. It enables very long coasts to be surveyed quickly, providing comprehensive overviews. These images allow coasts to be

Continued

mapped and changes monitored. Not only does it allow physical changes to be measured, but also pollution can be detected using ultraviolet wavelengths (Clark 1993). It has been particularly useful in tracking the development of oil spills, such as for the *Exxon Valdez* spill in Alaska (Stringer *et al.* 1992). A number of instruments have been introduced to monitor coasts, such as NASA's Coastal Zone Colour Scanner (CZCS). These data and information collected on the ground require storage, and GISs are rapidly becoming the main storage and retrieval system for such information. It allows layers of information to be superimposed, all of which can be interrogated by the user, and used to inform management decisions (O'Reagan 1996). Used together, remote sensing and GIS techniques are powerful tools in managing coastal problems (e.g. Populus *et al.* 1995).

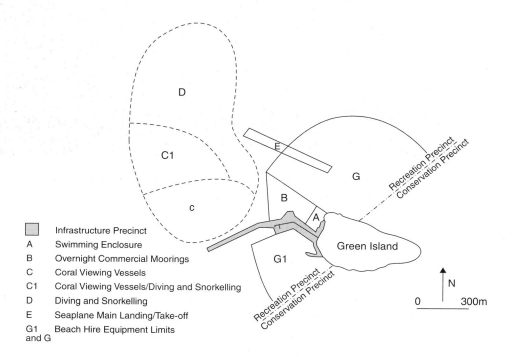

Figure 6.3 *The zoning plan for Green Island in the Cairns section of the Great Barrier Reef Marine Park (Australia).*
Source: Kay and Alder (1999) (Box 4.4, p. 125)

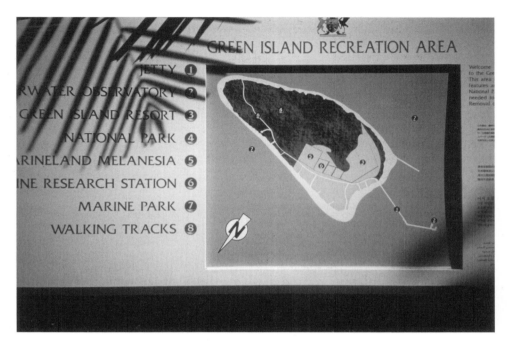

Plate 6.8 *Information for tourists on the management zoning of Green Island in the Cairns section of the Great Barrier Reef Marine Park.*

Summary

- Coastal management aims to facilitate sustainable use of the coastal zone and to protect human interests at the coast from natural and human-related processes.

- Coastal management is issue-led, and a number of key issues are identified, including pressure brought about through population growth, various uses of the coast, coastal hazards and administrative issues.

- Coastal management operates at many spatial scales, but to be fully effective small authorities should be integrated to provide overall consistent management along a stretch of coastline.

- Legislation is often required, but some non-statutory management schemes have proved successful.

Discussion questions

1 Citing at least one example, assess the impact of coastal population growth on the coastal environment.

2 Examine and prioritise the range of issues that should be addressed by the coastal manager.

3 Evaluate the benefits of integrated coastal zone management.

Further reading

See also

Definitions of the coastal zone, Chapter 1, Box 1.1
Waves and coastal management, Box 2.1.
Impact of tourism on coral cays, Box 2.4.
Human interference in bar-built estuaries, Box 3.3.
Deltas and human activity, Chapter 4.
Managing global sea level rise, Chapter 5.

General further reading

Coastal Planning and Management. R. Kay and J. Alder. 1999. E & FN Spon, London, 375 pp. A useful overview of planning and management strategies, provisioned with ample case studies from around the world.

Coastal and Estuarine Management. P. W. French. 1997. Routledge, London, 251 pp. Examines the impact of different activities on the coast.

The Geographical Journal, volume 164, part 3. R. J. Nicholls and J. Branson (eds). 1998. A special issue on coastal management in northwest Europe.

Studies in European Coastal Management. P. S. Jones, M. G. Healy and A. T. Williams. 1996. Samara Publishing Limited, Cardigan, 292 pp. A very useful set of papers on coastal management in a European context.

Glossary

abrasion   the wearing down of rock through frictional contact with rock fragments.

accretion   the addition of newly deposited sediment.

advection   the physical mixing of fluids of different density (e.g. fresh and salt water in an estuary).

afro-trailing edge coast   intermediate type of trailing edge coast.

allochthonous   material (sediment) transported to the site of deposition.

amero-trailing edge coast   mature type of trailing edge coast.

amino acid racemization   the alteration L-isoleucine to D-alloisoleucine amino acids in shell material following death. The L/D ratio is used as a relative dating technique.

amphidromic motion, point and system   the circular motion of a tidal wave around a fixed point due to the Coriolis effect.

Atlantic type   a coast where rock structure is perpendicular to the coastline.

attrition   reduction of rocks or rock fragments through erosional processes.

autochthonous   material (sediment) deposited *in situ*.

backshore   a morphological term describing the part of a beach that lies between mean high water and the landward limit of marine (storm wave) activity.

backwash   water returning to the sea under gravity following the breaking of a wave.

bar-built estuary   an estuary with a sedimentary feature, such as a bar or spit, partly obscuring the estuary mouth.

bay-head beach   an accumulation of sand that is trapped in an embayment by circulating currents, forming a beach.

beach cusp   rhythmical longshore scallop-shaped features on a beach, comprising an upstanding horn and a depressed embayment.

beachrock   a sand-sized deposit cemented by the precipitation of calcium carbonate. Occurs commonly in the tropics, but also in other areas, such as the Mediterranean.

beach zone   a sedimentological term for the area between mean low water and the landward limit of marine (storm wave) activity.

bedload transport   the movement of sediment along a surface near the bottom layer of the transporting fluid (air or water).

berm   a ridge situated relatively high up on a beach marking the limit of normal swash action.

bio-erosion   erosion caused by the activity of biological organisms.

bird's foot delta   a river-dominated delta with semi-radiating channels separated by water that appear like a bird's-footprint in aerial view.

black box system   a system where the inputs and outputs are known, but the pathway through the system is not.

blind estuary   an estuary whose connection to the sea is severed by the extension of a spit, for example, across the estuary mouth.

blowhole   a vertical hole in a cliff-top connected to a sea cave. Under certain conditions, water is forced out of the blowhole to produce a geyser.

blow-out   an erosion hollow in coastal dunes formed by localised deflation, often in response to vegetation undermining by animals or people.

bottomset beds   horizontally layered sediment beds deposited in front of a delta as it progrades seaward. They become covered and end up at the bottom of a stack of delta sediments.

braided river   a river comprising many interlaced channels.

braidplain   a plain supporting more than one braided river.

breaker zone   a process-based term for the area in the nearshore zone where incoming waves begin to break.

break-point   the point at which a wave begins to break during shoaling.

Bruun rule   a model suggesting the response of a sandy shoreline to sea level change is to retreat landward.

buoyancy-dominated river mouth   a deltaic river mouth where the river water is less dense than seawater and is, therefore, buoyant.

cascading system   an approach describing the movement of energy and material through a system.

cay   an island (sometimes vegetated) situated on a coral reef constructed from eroded coral debris and concentrated by wave refraction.

circulatory cell   refers to the circular pattern of currents and associated sediment movement in an enclosed coastal embayment.

clapotis   a standing wave that goes up and down, but does not appear to progress through the water.

closed system   a system where energy only transfers across the system boundaries.

coastal squeeze   the narrowing of the intertidal zone due to sea level rise in conjunction with permanent barriers (human-built or natural) preventing the landward migration of the intertidal zone.

coastal zone   the area between the seaward limit of terrestrial influence and the landward limit of marine influence.

constructive waves   these waves build up beach relief when long-period waves cause net-landward sediment transport, most commonly during fairweather wave conditions.

Coriolis effect   an effect whereby the spinning of the earth makes travelling objects appear to move to the right in the northern hemisphere and to the left in the southern hemisphere.

corrasion   the mechanical weathering of rock surfaces by abrasion and other related processes.

coseismic subsidence   subsidence of land coincident with an earthquake.

crevasse channel and splay   a channel formed by river water breaking through a levée and the splay of sediment deposited by such a breach.

cross-bedding   bedding planes in sediment that are at oblique angles to one another. Most often present in dune deposits where bedding planes representing both stoss- and lee-sides may be present.

cuspate foreland   in plan view, a triangular-shaped depositional landform protruding seaward of the coastline, often arising where there is an abundant sediment supply and where two opposing longshore drift systems meet, piling up sediment into this example of an equilibrium trap. The point of the triangular cuspate foreland usually faces the direction of minimum fetch.

deep sink   any store of sediment below the influence of waves and onshore currents. Sediment that finds its way into a deep sink is lost from the coastal system until reworked under different sea levels.

deflation   the erosion and removal of sand by aeolian processes.

destructive waves   waves that lower beach relief under short-period wave conditions, usually associated with storm wave conditions.

diatom   unicellular algae that inhabit all aquatic environments. They possess a frustule that is composed of silica that are often preserved as microfossils in coastal sediments and useful in palaeoenvironmental studies.

differential erosion   uneven erosion of a coastline due to local variations in the properties (e.g. hardness, structure) of adjacent rock-types.

diffraction   a process whereby waves can enter wave shadow zones (e.g. in the lee of an island or other obstacle) due to gravity allowing water to propagate or 'bleed' laterally out of the side of a wave 'severed' by the obstacle.

diffusion   the mixing of fluids of different ionic compositions in order to obtain uniformity (e.g. fresh and salt water in an estuary).

dispersion   the separation of different wave trains, through variations in wave travel time, i.e. long wave-length waves travel faster than short wave-length waves.

dissipative domain   a beach which dissipates and absorbs wave energy.

distributary channels   river channels on a delta that commonly diverge as they approach the shore.

diurnal tide   a tidal cycle possessing one high and one low tide per day.

dominant current   refers to the dominant current in a tidal environment, i.e. flood or ebb tide currents.

drift-aligned   a coast that is aligned obliquely to prevailing incident waves.

dune slack   a depression in a coastal dune system that often separates dune ridges.

dynamic equilibrium   a state of equilibrium where environmental conditions fluctuate around an average that is itself gradually changing.

ebb tide   outgoing tidal flow.

ecosystem   the interaction of organisms with the physical environment.

edge waves   reflected waves that travel in a direction quasi-parallel to the shore and interact with incident waves.

embryo dune   a mound of wind-blown sand accumulating around the high water level and colonised by pioneer dune vegetation. Embryo dunes often develop into foredunes.

endogenetic processes   internal earth-processes (e.g. plate tectonics) driven by geothermal energy.

equilibrium trap   a coastal sediment trap commonly occurring where two opposing longshore drift systems meet. Examples include cuspate forelands and tombolos.

estuary mouth bypassing   the offshore removal of estuarine/deltaic sediment by ebb-tidal currents

eustasy   absolute level of the sea surface and its fluctuations.

eutrophication   the stagnation of an aquatic ecosystem through oxygen-depletion brought about by high nutrient levels that over-stimulate biological production.

exogenetic processes   processes operating on the earth's surface, driven by solar energy.

facies   the characteristics of a sediment that represents the environment of deposition.

fairweather wave   long period, low height waves that usually characterise summer wave conditions.

fan deltas   deltas characterised by high levels of fluvial energy and consequent coarse-grained sediment deposition (there is a number of subtypes explained in the main text).

feeder system   the general term for a fan delta's fluvial sediment source.

fetch   the distance over which waves travel.

fish-hook beaches   beaches that are separated by headlands and become wider down-drift between the headlands.

fjord   a glaciated valley drowned by sea level rise.

floc   a relatively large particle formed by the coagulation of relatively small individual clay particles during the process of flocculation

flocculation   the coagulation of clay particles of river origin upon encountering salt nucleii in an estuary.

flood tide   incoming tidal flow.

flood tide delta   a deltaic accumulation of sediment landward of the mouth of an estuary, deposited by dominant flood tide currents.

fluid threshold velocity   velocity of a fluid required to initiate movement of sediment of a given particle size.

forced waves   waves that are actively being generated.

foredune   the most seaward ridge of a coastal dune system.

foreset beds   seaward-sloping sediment beds deposited at the delta front.

foreshore   a morphological term for the area on a beach between mean low and high water.

free dunes   sand dunes that are not colonised by vegetation and so free to move in response to changing wind conditions.

free waves   waves that have left the area in which they were generated.

friction-dominated river mouth   shallow delta river mouths where friction with bed forces the emerging flow to spread laterally along the coast.

fringing reefs   coral reefs that fringe the mainland or continental islands.

geoid   the uneven surface of the sea, by up to 100 m, due to gravity differences in the earth.

Gilbert-type delta   a delta comprising characteristic bottomset, foreset and topset beds.

glacio-eustasy   changes in absolute sea level brought about by the formation of continental ice, extracting oceanic water during an ice age.

glacio-isostasy   changes in land level brought about by the growth and decay of continental ice. The weight of ice pushes the land down, rebounding on deglaciation. This influences regional relative sea levels.

grey box system   a system where input and outputs are known, along with some appreciation of the internal workings of the system.

groyne   an artificial structure positioned across a beach at right angles to the shore to interrupt and trap sediment being transported by longshore currents.

H   abbreviation of wave-height.

H_s   abbreviation of significant wave-height.

halocline   level in the water column where salinity abruptly changes, i.e. the contact between fresh surface and deep saline waters in an estuary.

halophytes   plants able to tolerate saline conditions.

Holocene epoch   from 10 ka to the present day (current interglacial).

homogeneous estuary   an estuary with uniform salinity throughout.

homopycnal flow   where river water and the receiving waterbody possess equal densities.

hydraulic pressure   pressure created in a rock void through the trapping of air or water by breaking waves, serving to weaken rock.

hydrological cycle   describes the transfer of water between natural storages.

hyperpycnal flow   where river water is received by a waterbody of lower density.

hypersynchronous estuary   where tidal range increases up-estuary.

hypopycnal flow   where river water enters a waterbody of higher density.

hyposynchronous estuary   where tidal range decreases up-estuary.

impact threshold velocity   velocity required to move sediment grains of a given particle size through the combined influence of the fluid velocity and energy transferred to the sediment grain by collision with other grains already in motion.

impeded dunes   sand dunes that are semi-fixed by vegetation colonisation.

incident wave   a wave approaching a shore, usually for the first time, i.e. not a reflected wave.

indicative meaning and range   the position of a sea level index point in the tidal frame at the time of formation/deposition.

inertia-dominated river mouth   river outflow in a delta setting that is unconfined, spreading out in all directions forming a lobate delta.

infauna   animals living within sediment.

inner shelf delta   a delta situated in shallow water in an inner shelf position.

inshore   a morphological term for the area between wave-base and mean low water.

interdistributary areas   deltaic areas lying between distributary channels.

interseismic strain accumulation   strain accumulated along a fault between earthquake events, often resulting in minor uplift.

intertidal zone   the area lying between high and low tide levels.

isolation basin   a coastal basin cut off and isolated from the sea by relative sea level fall.

isostasy   the vertical movement of the land due to the loading and unloading of weights (e.g. ice).

karst   features associated with the chemical dissolution of limestone, such as cave systems.

λ   abbreviation of wave-length.

lagoonal estuary   an estuary whose connection to the sea is severed by the extension of a spit, for example, across the estuary mouth.

laterally homogeneous estuary   salinity is uniform across, but changes along, an estuary.

laterally inhomogeneous estuary   salinity changes across and along an estuary, but is uniform at all depths at any given point.

leading edge coast   a coast occurring on a continental margin close to a converging crustal plate boundary.

level of no motion   a level in the water column of an estuary separating opposing upper and lower water layers where no tidal flow occurs.

lithology   rock-type.

littoral energy fence   onshore wave energy that prevents sediment escaping offshore.

longshore drift   currents that move along a shore, often transporting sediment.

lutocline   the gradient of suspended sediment concentration in the water column of an estuary.

macrotidal   a tidal environment with a tidal range > 4 m.

managed retreat   allowing and/or managing the landward retreat of a shoreline due to coastal erosion and/or sea level rise.

marginal sea coast   a coast on a continental margin some distance from a converging crustal plate boundary.

mesotidal   a tidal environment with a tidal range 2–4 m.

meta-stable equilibrium   a system that switches/jumps from one level of equilibrium to another due to trigger events.

microtidal   a tidal environment with a tidal range < 2 m.

mixed tide   a tidal cycle comprising one dominant high and low tide per day, but with additional secondary (significantly smaller) high and low tides.

morphodynamic   changing morphology in response to variations in environmental conditions.

morphological system   an approach that deals with the morphological expression of a system.

neap tide   tide that occurs when the sun and moon are perpendicular as regards the earth, to produce lower than average high tides and higher than average low tides.

nearshore zone   process-based term for the area comprising swash, surf and breaker zones.

negative feedback   a response to a system change that aids recovery of the original state of equilibrium.

neo-trailing edge coast   immature type of trailing edge coast.

non-turbid estuary   an estuary with sediment too coarse for it to be held in suspension for long periods of time, so that the water is relatively clear.

null point   the point on an estuary bed where opposing flows meet, reducing flow velocity to zero.

offshore   a morphological term for the area below wave-base.

offshore-transition zone   the area between fairweather and storm wave-bases.

offshore zone   a sedimentological term for the area lying below storm wave-base.

open-ended estuary   an estuary with an unobscured connection to the sea.

open system   a system where energy and matter are transferred across its boundaries.

orthogonals lines drawn at right angles to a wave-front and extended to the shore, indicating wave-energy concentration received along any given length of the coast, i.e. where orthogonals converge wave-energy is concentrated, but is dissipated where orthogonals diverge.

oscillatory wave a deep-water wave in which water particles move in an orbital motion.

Pacific type a coast where the rock structure is parallel to the coastline.

parabolic dune a crescentic impeded dune commonly found on the downwind perimeter of blow-outs, with its arms opening upwind.

parallel orientated bedforms sedimentary bedforms aligned parallel with the prevailing current direction.

partially-closed estuary an estuary with a sedimentary feature, such as a bar or spit, partly obscuring the estuary mouth.

particle size refers to the size of both individual sediment particles or entire populations.

patch reef subrounded platform-like coral reefs located on the continental shelf.

permeability the ability of a rock-type to allow water to pass through it via structural and bedding planes.

perpendicular orientated bedforms sedimentary bedforms aligned perpendicular with the prevailing current direction.

Pleistocene epoch the first epoch of the Quaternary, from 1.81 Ma to 10 ka.

plunging breaker breaking waves that characterise steeply inclined shores and comprise a steep-fronted wave that curls over, crashing down on the shore producing foam.

plunging cliffs quasi-vertical cliffs that extend below sea level. Commonly found on previously glaciated fjordic coasts where U-shaped valleys have been drowned with the steep sides forming plunging cliffs.

pocket beach a small beach located between headlands.

positive feedback a response to a system change that enhances the change leading to a new state of equilibrium.

pressure-release jointing the removal of rock from a cliff front releases confining pressure that manifests itself by the creation of cliff-parallel joints.

primary dune coastal sand dunes which are orientated parallel to the shore in their original depositional position.

process-response system an approach which recognises that morphology is determined through response to processes.

progradation   the seaward advance of a sedimentary shoreline, usually through excess sediment input over output.

quarrying   the detachment and removal of rock fragments from cliffs and shore platforms by wave activity.

Quaternary   the most recent geological period from 1.81 million years ago to the present. Comprises the Pleistocene and Holocene epochs and is characterised by alternating glacial and interglacial conditions.

radiocarbon dating   dating of carbonaceous material using the decay and half-life of ^{14}C.

raised beach   a fossil beach that occurs above present-day sea level.

re-entrant trap   a sediment trap, such as a bayhead beach or pocket beach, where sediment introduced into an embayment is unable to escape due to a circulatory cell operating within the embayment.

reflective domain   a steeply inclined beach state that reflects wave energy back out to sea.

refraction   the bending of a wave front caused by a velocity reduction when the wave enters shallow water, so that the seafloor is equal to or shallower than wave-base.

regression   sea level fall and the emergence of coastal land.

relative sea level change   the level of the sea relative to the land, determined by eustasy and isostasy.

relief   character of the landsurface, including topography and elevation. High relief refers to mountainous or hilly terrain, whilst low relief refers to low and rather featureless terrain.

retrogradation   the landward retreat of a sedimentary shoreline, usually through a deficit of sediment input over output.

ria   a river valley drowned by sea level rise.

ribbon reef   a linear coral reef orientated parallel to the shore and usually located along the seaward edge of the continental shelf.

rip-current   a seaward directed current associated with water returning to the sea after being brought onshore by wave-breaking activity.

river mouth bypassing   sediment flushed out to sea by high river discharge (e.g. flood event).

rollover   the gradual net landward migration of a subaerial sediment barrier, mainly refers to gravel barriers.

runnels   shore-parallel gullies occurring on the low-angled facet of a beach near the low water level and separated by ridges, although may be connected through the ridges by rip-channels.

sabka   intertidal to supratidal salt flats that occur along arid coastlines, often characterised by algal mats, salt crusts and other evaporite minerals.

salient trap   a sediment trap that extends outwards from the shore, such as a spit.

salinisation   the increase of salt levels in soil due to high evaporation rates or saline groundwater.

saltation   a mode of sediment transport that involves sediment grains moving in a series of jumps or hops.

salt-balance principle   a model proposed by Pritchard (1955) to explain the mixing of fresh and salt water in an estuary, through the processes of advection and diffusion, with implications for the classification of estuaries.

salt pan   a pool of saline water found on a salt marsh surface.

salt-wedge   a distinct wedge of dense saline water that occurs at the bottom of an estuarine water column and tapers upstream. Common in estuaries where fresh–salt watermixing is limited.

sea level index points   points used to construct sea level curves for which age, altitude, indicative meaning and range and tendency of sea level are known.

sea waves   wind-generated waves encountered within the generation area, often high with short period.

secondary dune   a coastal dune that has been reworked by coastal processes to form a ridge at ridge-angles to the shore.

sediment couplet   alternating coarse and fine sediment layers, often associated with tidal environments where the coarse sediment is deposited by the dominant current and the fine sediment by the subordinate current.

sediment sorting   similarity of sediment populations in terms of either size, shape or both. A sediment comprising grains of a similar size and/or shape is said to be well-sorted.

seiche   a type of standing wave, created by incident wave reflection, that sloshes about in enclosed coastal waters. Severe seiches in harbours may lift boats out of the water.

semi-diurnal tide   a tidal cycle comprising two high and two low tides per day.

settling lag   the phenomenon whereby sediment particles can undergo net shoreward transport in a tidal environment.

settling velocity   the velocity at which a sediment particle may settle out of the water column.

shadow dune   an unvegetated accumulation of wind-blown sand in the lee of an obstacle, often at the strandline where marine debris is plentiful.

shelf edge delta   a delta built up at a site on the edge of a continental shelf through abundant sediment deposition outpacing the rise of sea level.

shoaling   occurs when the wave-base of a deep-water oscillatory wave first intersects the sea-floor, so beginning the transition to a translatory wave, involving the formation of breaking waves.

shoreface bypassing   offshore removal of coastal sediment by storm-induced offshore bottom currents.

shoreface zone   a sedimentological term for the area between mean low water and fairweather wave-base.

shore platform   an eroded surface in the intertidal zone. Often referred to as wave-cut, but other processes such as salt weathering and bio-erosion are also important.

significant wave-height (H$_s$)   the mean of the highest one-third of waves encountered on a coast.

spilling breaker   a type of breaking wave that is characteristic of gently sloping beaches. Producing lots of foam.

spit   a depositional landform that occurs where the shore orientation changes, but longshore currents do not deviate and continue to transport and deposit along a projected coastline.

spring tide   a tide that occurs when the sun and moon are aligned as regards the earth to produce higher than average high tides and lower than average low tides.

standing wave   a wave that goes up and down, but does not appear to progress.

steady state equilibrium   where environmental conditions vary around an unchanging average state.

steric sea level change   change due to ocean water density changes, e.g. warm water is less dense, expands causing sea level to rise.

stillstand   a period of time over which sea level does not change.

stillwater level   mean water level, half way between the depth of a wave's trough and the height of the crest.

storm profile   a beach profile developed under storm wave conditions where destructive waves lower the beach gradient.

storm surge   coastal inundation of sea water associated with storm conditions, especially high onshore winds that pile water up at the coast, and the effect of low pressure that allows sea level to rise by 1 cm for every mb of pressure change.

submerged forest   woodland that has been drowned by sea level rise, preserved in peat, and subsequently exposed in intertidal and subtidal zones.

subordinate current   refers to the least effective current in a tidal environment, i.e. flood or ebb tide currents.

summer profile   a beach profile developed under fairweather wave conditions where constructive waves build up the beach gradient.

superposition   the merging of two or more wave trains travelling in the same direction. Where the different waves are in phase (i.e. the crest and troughs of each wave train are superimposed), the wave amplitude is increased, whereas amplitude is decreased where the wave trains are not in phase. Gives rise to the phenomena known as surf beat.

surface creep   a mode of sediment transport which involves sediment grains being moved along a surface.

surf beat   the alternating arrival of sets of higher and lower than average wave amplitudes on the coast. See **superposition**.

surf zone   a process-based term for the area of the nearshore zone where breaking waves approach the shore, usually over a wide low-gradient surface.

surging breaker   a type of breaking wave that surges up a steep beach by forcing water out the front of the wave, producing little foam.

suspended particulate matter (SPM)   particles held in suspension in an estuary.

swash   the movement of water up the beach upon a wave breaking.

swash-aligned   the orientation of a shoreline parallel to approaching wave fronts.

swash zone   process-based term for the area of the nearshore zone where waves finally break and travel up the beach as swash. Also, water returns to the sea as backwash within this zone.

swell profile   a beach profile developed under fairweather wave conditions where constructive waves build up the beach gradient.

swell waves   wind-generated waves encountered outside the area of generation, often of low height and long period.

synchronous estuary   where tidal range is uniform throughout an estuary.

T   see **wave-period**.

tendency of sea level   the direction of sea level change indicated by a sea level index point. Positive/negative tendency refers to sea level rise/fall.

thermoluminescence dating   dating minerals based on trapped thermoluminescent energy since burial. Exposure to light bleaches the minerals and resets the thermoluminescent clock back to zero.

tidal bore   a wave produced in a hypersynchronous estuary at the onset of the flood tide, where the estuary narrows, shallows and the tidal range increases upstream.

tidal bulge   bulge of water pulled up by the gravitational attraction of the moon and the sun.

tidal creek   creek which facilitates the flooding and withdrawal of tidal waters on and off a mudflat, salt marsh or mangrove surface.

tidal currents   currents produced by the incoming and outgoing of tidal water at the coast; particularly effective in constricted waterways, such as estuaries.

tidal cycle   the rise and fall of the tide as predicted by gravitational effects of the moon and sun on the sea surface.

tidal prism   the amount of water that enters and exits an estuary every flood and ebb tide respectively.

tidal range   the vertical distance between high and low tide at a given location along the coast.

tidal rhythmites   sediment couplets that reflect the rhythm of the tides, e.g. thicker sediment layers during spring tides and thinner during neap tides.

tide-dominated estuary   an estuary whose processes are dominated by tides.

tombolo   an accumulation of sediment extending from the shore to an offshore island or sandbank.

toppling failure   a mechanism of cliff erosion and retreat whereby a stack of rock topples forward, with the base of the stack remaining approximately in its original position, like a tree being felled.

topset beds   horizontal sediment beds deposited on a delta plain.

trailing edge coast   a coast formed by crustal divergence with subsequent drift away from the plate boundary – tectonically stable.

transgression   the rise of sea level and coastal indundation.

translatory wave   a shallow-water wave in which water particles move in a highly elliptical motion.

tsunami   waves created by submarine earthquakes, slides, volcanic activity and meteor/comet impacts in the ocean. Often tsunami can be catastrophic, particularly around the Pacific rim, and may make significant contributions to the development of coastlines in these areas.

turbid estuary   an estuary with sediment fine enough for it to be held in suspension for long periods of time, so that the water appears muddy.

turbidity   refers to the amount of sediment held in suspension by a fluid. Clear water is non-turbid, whilst visibly sediment-rich fluid has high turbidity.

turbidity maximum   a zone in the waters of an estuary where maximum suspended sediment concentration occurs.

uranium-series dating   a dating technique for mineral material that uses the decay and half-life of uranium, thorium, radon and lead.

wave-base   equates to a water depth below which passing waves have no effect (variably defined as being equal to between ½ and ¼ of the wave-length).

wave-cut notch   an indentation at the base of a cliff excavated by wave processes. Often leads to undermining of a cliff and subsequent cliff collapse.

wave-dominated estuary   an estuary whose processes are dominated by wave activity.

wave-height (H)   the vertical distance between the wave trough and crest.

wave-length (λ)   the horizontal distance between consecutive wave crests or troughs.

wave-period (T)   the time interval between consecutive wave crests or troughs passing a fixed point.

wet slack   a depression between coastal dune ridges where the surface intersects the water table creating wet conditions.

white box system   a system for which the inputs, ouptuts and detailed workings are known.

winter profile   a beach profile developed under storm wave conditions where destructive waves lower the beach gradient.

xerophytic   a plant able to withstand harsh environmental conditions, such as drought. Many plant species growing on coastal sand dunes, where water is scarce, are xerophytic.

zeta-form beach   beaches that are separated by headlands and become wider down-drift between the headlands.

● Further reading

General

Coasts: An Introduction to Coastal Geomorphology (3rd Edition). E. C. F. Bird. 1984. Blackwell, Oxford, 320 pp. A standard introduction to coastlines.

Coastal Environments: An Introduction to the Physical, Ecological and Cultural Systems of Coastlines. R. W. G. Carter. 1988. Academic Press, London, 617 pp. A thorough advanced-level overview of the coast, quite technical in places and becoming out-of-date.

The Evolving Coast. R. A. Davis, Jr. 1994. Scientific American Library, New York, 231 pp. A concise introduction to the physical development of coastlines.

Coastal and Estuarine Management. P. W. French. 1997. Routledge, London, 251 pp. A timely introduction to various management issues currently affecting coastal systems.

Coasts. J. D. Hansom. 1988. Cambridge University Press, Cambridge, 96 pp. A concise and readable introduction to coasts.

Les littoraux: Impact des aménagements sur leur évolution. R. Paskoff. 1998. Armand Collin, Paris, 260 pp. An excellent French language text dealing with coastal problems.

Introduction to Coastal Geomorphology. J. Pethick. 1984. Arnold, London, 260 pp. Although titled 'Introduction' this is in fact a fairly technical, and now slightly out-dated, guide to coastal geomorphology, but highly recommended for Pethick's elegant writing and excellent diagrams.

Exploring Ocean Science (2nd Edition). K. Stowe. 1996. Wiley, New York, 426 pp. A very well-illustrated and thorough review of oceanography. Only some of the chapters deal with the coastal ocean, but covers processes, landforms and ecology.

Coastal Dynamics and Landforms. A. S. Trenhaile. 1997. Oxford University Press, Oxford, 365 pp. An advanced-level text dealing with process and form at the coast.

Coastal Problems: Geomorphology, Ecology and Society at the Coast. H. Viles and T. Spencer. 1995. Arnold, London, 350 pp. A readable text on problems affecting the global coastline.

The periodicals *Journal of Coastal Research, Estuarine, Coastal and Shelf Science, Shore and Beach, Journal of Coastal Conservation, Marine Geology* and others should be consulted regularly for individual research papers, themed sections, and special issues devoted to particular topics.

Specific topics

Waves, tides and currents

Waves, Tides and Shallow-water Processes. Open University. 1989. Pergamon Press, Oxford, 187 pp. A thorough text focusing on the operation of coastal processes.
Tides, Surges and Mean Sea-Level. D. T. Pugh. 1987. Wiley, Chichester, 472 pp.
Tsunamis in the World. S. Tinti (ed.). 1992. Kluwer Academic, Dordrecht, 228 pp.

Rocky coasts

The Biology of Rocky Shores. C. Little and J. A. Kitching. 1996. Oxford University Press, Oxford, 240 pp.
Geomorphology of Rocky Coasts. T. Sunamura. 1992. Wiley, Chichester, 302 pp. An excellent and technical account of erosional processes and resultant landforms at the coast.
The Geomorphology of Rock Coasts. A. S. Trenhaile. 1987. Oxford University Press, Oxford, 384 pp. A thorough advanced-level overview of processes and landforms of erosional coastlines.
Intertidal Ecology. D. Raffaelli and S. Hawkins. 1996. Chapman and Hall, London, 356 pp. A timely general introduction to the ecology of intertidal environments along all types of coasts.

Coral reefs

Coral Reef Geomorphology. A. Guilcher. 1988. Wiley, Chichester, 228 pp. An introduction to the physiology of coral reef systems throughout the world.
The Geomorphology of the Great Barrier Reef: Quaternary Development of Coral Reefs. D. Hopley. 1982. Wiley, New York, 453 pp. An exhaustive account of one of the world's most spectacular coastal environments.
Discover the Great Barrier Reef Marine Park (Revised Edition). L. Murdoch (compiler). 1996. Angus & Robertson, Sydney, 96 pp. A very well-illustrated and accessible introduction to the physical setting, ecology, history, conservation and management of Australia's Great Barrier Reef.
Ecosystems of the World (Volume 25): Coral Reefs. Z. Dubinsky (ed.). 1990. Elsevier, Amsterdam.
Oceanic Islands. P. D. Nunn. 1994. Blackwell, Oxford.

Beaches, barrier islands and coastal dunes

Beaches: Form and Process. J. Hardisty. 1990. Unwin Hyman, London, 324 pp. An advanced-level text introducing beach and nearshore dynamics.

'Beach and nearshore sediment transport'. J. Hardisty. 1994. In *Sediment Transport and Depositional Processes* K. Pye (ed.). Blackwell Scientific Publications, Oxford, pp. 219–255.

Coastal Dunes: Form and Process. K. F. Nordstrom, N. Psuty and B. Carter (eds). 1990. Wiley, Chichester, 392 pp. A collection of papers dealing with the development and geomorphology of coastal dunes with good case study material.

Coastal Dune Management. D. Ranwell and R. Boar. 1986. Institute of Terrestrial Ecology, Huntingdon, 105 pp.

Marine Biology: Function, Biodiversity and Ecology. J. S. Levington. 1995. Oxford University Press, Oxford. Recommended for its treatment of beach ecosystems.

Tidal environments

Estuaries: A Physical Introduction (2nd Edition). K. R. Dyer. 1997. Wiley, Chichester, 210 pp. An introductory text to estuaries, including their classification, tidal effects and mixing processes, but quite mathematical in places.

Estuarine Shores: Evolution, Environments and Human Alterations. K. F. Nordstrom and C. T. Roman (eds). 1996. Wiley, New York, 510 pp. An advanced compilation of work on the very varied nature of estuarine environments.

The Estuarine Ecosystem (2nd Edition). D. S. McLusky. 1989. Blackie, Glasgow, 215 pp. A now standard and well-accepted introduction to estuarine ecosystems.

Saltmarshes: Morphodynamics, Conservation and Engineering Significance. J. R. L. Allen and K. Pye (eds). 1992. Cambridge University Press, Cambridge. A useful and wide-ranging introduction to salt marsh systems, their ecology and applied issues.

Intertidal Ecology. D. Raffaelli and S. Hawkins. 1996. Chapman and Hall, London, 356 pp. A timely general introduction to the ecology of intertidal environments along all types of coasts.

Mangrove – The Forgotten Habitat. J. S. Deitch. 1996. Immel Publishing, London.

The Botany of Mangroves. P. B. Tomlinson. 1986. Cambridge University Press, Cambridge.

Deltaic environments

'Clastic coasts'. H. G. Reading and J. D. Collinson. 1996. In: H. G. Reading (ed.) *Sedimentary Environments: Processes, Facies and Stratigraphy*. Blackwell, Oxford, 154–231. A substantial chapter on modern and ancient clastic coasts, but with a clear emphasis on deltaic sedimentary environments.

Journal of Coastal Research, volume 14, part 3. 1998. A timely thematic issue devoted to deltaic coastal environments, with some excellent review articles, including the Mississippi Delta and Arctic deltas, as well as research papers. A good source of recent delta literature.

Coarse-grained Deltas. A. Colella and D. B. Prior (eds). 1990. Blackwell Scientific Publications, Oxford, 357 pp. A compilation of research papers on fan deltas.

Deltas: Sites and Traps for Fossil Fuels. M. K. G. Whateley and K. T. Pickering (eds). 1989. Geological Society, London, Special Publication No. 41, 360 pp. As the title suggests, this is a collection of geological research papers devoted to hydrocarbon exploration in deltaic regions.

Records of sea level change

Sea-Level Changes: The Last 20,000 Years. P. A. Pirazzoli. 1996. Wiley, Chichester, 211 pp. A well-written introduction to sea level changes on a geological timescale.

Ice Age Earth: Late Quaternary Geology and Climate. A. G. Dawson. 1992. Routledge, 293 pp. An extensive review of climate and environmental change from the last glacial onwards.

Sea-level Research: A Manual for the Collection and Evaluation of Data. O. van de Plassche (ed.). 1986. Geo Books, Norwich, 618 pp. The standard reference work for the collection and interpretation of sea level data.

Impacts of sea level rise

Submerging Coasts: The Effects of a Rising Sea Level on Coastal Environments. E. C. F. Bird. 1993. Wiley, Chichester, 184 pp. A useful review of the response of coastlines to rising sea levels.

Climate and Sea-level Change: Observations, Projections, and Implications. R. A. Warrick, E. M. Barron, and T. M. L. Wigley (eds.). 1993. Cambridge University Press, Cambridge, 424 pp. A collection of papers examining the issue of global climate change and associated sea level rise.

Impacts of Sea-level Rise on European Coastal Lowlands. S. Jelgersma and M. J. Tooley (eds). 1992. Blackwell, Oxford, 267 pp. An example of how sea level rise impacts on a regional scale.

Coastal management

Coastal Planning and Management. R. Kay and J. Alder. 1999. E & FN Spon, London, 375 pp. A useful overview of planning and management strategies, provisioned with ample case studies from around the world.

Coastal and Estuarine Management. P. W. French. 1997. Routledge, London, 251 pp. Examines the impact of different activities on the coast.

The Geographical Journal, volume 164, part 3. R. J. Nicholls and J. Branson (eds). 1998. A special issue on coastal management in northwest Europe.

Studies in European Coastal Management. P. S. Jones, M. G. Healy and A. T. Williams. 1996. Samara Publishing Limited, Cardigan, 292 pp. A very useful set of papers on coastal management in a European context.

Directions in European Coastal Management. M. G. Healy and J. P. Doody. 1995. Samara Publishing Limited, Cardigan, 556 pp. The conference proceedings of Coastlines '95, contains a number of very useful papers.

Bibliography

Allen, J. R. L. 1989. 'Evolution of salt-marsh cliffs in muddy and sandy systems: a qualitative comparison of British west-coast estuaries'. *Earth Surface Processes and Landforms*, 14, 85–92.

Allen, J. R. L., 1993. 'Muddy alluvial coasts of Britain: field criteria for shoreline position and movement in the recent past'. *Proceedings of the Geologists' Association*, 104, 241–262.

Allen, J. R. L., 1994. 'A continuity-based sedimentological model for temperate-zone tidal salt marshes'. *Journal of the Geological Society, London*, 151, 41–49.

Allen, J. R. L., 1996. 'Shoreline movement and vertical textural patterns in salt marsh deposits: implications of a simple model for flow and sedimentation over tidal marshes'. *Proceedings of the Geologists' Association*, 107, 15–23.

Allen, J. R. L., 1999. 'Geological impacts on coastal wetland landscapes: some general effects of sediment autocompaction in the Holocene of northwest Europe'. *The Holocene*, 9, 1–12.

Alve, E. 1995. 'Benthic foraminiferal responses to estuarine pollution: a review'. *Journal of Foraminiferal Research*, 25, 190–203.

Anthony, E. J. and Blivi, A. B., 1999. 'Morphosedimentary evolution of a delta-sourced, drift-aligned sand barrier-lagoon complex, western Bight of Benin'. *Marine Geology*, 158, 161–176.

Aslan, A. and Autin, W. J., 1999. 'Evolution of the Holocene Mississippi River floodplain, Ferriday, Louisiana: insights on the origin of fine-grained floodplains'. *Journal of Sedimentary Research*, B69, 800–815.

Bandyopadhyay, S., 1997. 'Natural environmental hazards and their management: a case study of Sagar Island, India'. *Singapore Journal of Tropical Geography*, 18, 20–45.

Bascom, W. H., 1951. 'The relationship between sand size and beach slope'. *Transactions of the American Geophysical Union*, 32, 866–874.

Bayliss-Smith, T. P., Healey, R., Lailey, R., Spencer, T. and Stoddart, D. R., 1979. 'Tidal flows in salt marsh creeks'. *Estuarine and Coastal Marine Science*, 9, 235–255.

Bijlsma, L., 1996. 'Coastal zones and small islands'. In: R. T. Watson, M. C. Zinyowera and R. H. Moss (eds) *Climate Change 1995: Impacts, Adaptations and Mitigation of Climate Change: Scientific-Technical Analyses (Contribution of Working Group II to the Second Assessment Report of the Intergovernmental Panel on Climate Change)*. Cambridge University Press, Cambridge, 289–324.

Bishop, P. and Cowell, P., 1997. 'Lithological and drainage network determinants of the character of drowned, embayed coastlines'. *Journal of Geology*, 105, 685–699.

Boomer, I., 1998. 'The relationship between meiofauna (ostracoda, foraminifera) and tidal levels in modern intertidal environments of north Norfolk: a tool for palaeoenvironmental reconstruction'. *Bulletin of the Geological Society of Norfolk*, 46, 17–26.

Boyd, R., Dalrymple, R. W. and Zaitlin, B. A., 1992. 'Classification of clastic coastal depositional environments'. *Sedimentary Geology*, 80, 139–150.

Bray, M., Hooke, J. and Carter, D., 1997. 'Planning for sea-level rise on the south coast of England: advising the decision-makers'. *Transactions of the Institute of British Geographers*, 22, 13–30.

Briggs, D., Smithson, P., Addison, K. and Atkinson, K., 1997. *Fundamentals of the Physical Environment* (2nd Edition). Routledge, London.

Broche, P., Devenon, J. L., Forget, P., de Maistre, J. C., Naudin, J. J. and Cauwet, G., 1998. 'Experimental study of the Rhône plume. Part I: physics and dynamics'. *Oceanologica Acta*, 21, 725–738.

Brown, B. E. and Ogden, J. C., 1993. 'Coral bleaching'. *Scientific American*, 268, 44–70.

Bruun, P., 1962. 'Sea level rise as a cause of shore erosion'. *Journal of Waterways and Harbour Division*, 88, 117–130.

Bryant, E. A. and Price, D. M., 1997. 'Late Pleistocene marine chronology of the Gippsland Lakes region, Australia'. *Physical Geography*, 18, 318–334.

Bryant, E. A., Young, R. W. and Price, D. M., 1996. 'Tsunami as a major control on coastal evolution, southern Australia'. *Journal of Coastal Research*, 12, 831–840.

Bryant, E. A., Young, R. W. and Price, D. M., 1997. 'Late Pleistocene marine deposition and TL chronology of the New South Wales, Australian coastline'. *Zeitschrift für Geomorphologie N. F.*, 41, 205–227.

Budd, W. F. and Smith, I. N., 1985. 'The state of balance of the Antarctic ice sheet – an updated assessment 1984'. In: *Glaciers, Ice Sheets and Sea Level: Effects of a CO_2-induced Climatic Change*. National Academy Press, Washington, 172–177.

Carriquiry, J. D. and Sanchez, A., 1999. 'Sedimentation in the Colorado River Delta and upper Gulf of California after nearly a century of discharge loss'. *Marine Geology*, 158, 125–145.

Carter, R. W. G., 1988. *Coastal Environments: An Introduction to the Physical, Ecological and Cultural Systems of Coastlines*. Academic Press, London.

Carter, R. W. G. and Orford, J. D., 1993. 'The morphodynamics of coarse clastic beaches and barriers: a short- and long-term perspective'. *Journal of Coastal Research, Special Issue*, 15, 158–179.

Carter, R. W. G. and Stone, G. W., 1989. 'Mechanisms associated with the erosion of sand dune cliffs, Magilligan, Northern Ireland'. *Earth Surface Processes and Landforms*, 14, 1–10.

Charman, D. J., Roe, H. M. and Gehrels, W. R., 1998. 'The use of testate amoebae in studies of sea-level change: a case study from the Taff Estuary, South Wales, UK'. *The Holocene*, 8, 209–218.

Chen, X. and Zong, Y., 1999. 'Major impacts of sea-level rise on agriculture in the Yangtze delta area around Shanghai'. *Applied Geography*, 19, 69–84.

Cicin-Sain, B., 1993. 'Sustainable development and integrated coastal zone management'. *Ocean and Coastal Management*, 21, 11–44.

Clark, C., 1993. 'Satellite remote sensing of marine pollution'. *International Journal of Remote Sensing*, 14, 2985–3004.

Cremona, J., 1988. *A Field Atlas of the Seashore*. Cambridge University Press, Cambridge.

Dalrymple, R. W., Zaitlin, B. A. and Boyd, R., 1992. 'A conceptual model of estuarine sedimentation'. *Journal of Sedimentary Petrology*, 62, 1130–1146.

Davies, J. L., 1964. 'A morphogenic approach to the world's shorelines'. *Zeitschrift für Geomorphologie*, 8, 127–142.

Davies, K. H., 1983. 'Amino acid analysis of Pleistocene marine molluscs from the Gower Peninsula'. *Nature*, 302, 137–139.

Davis, R. A., Jr., 1994. *The Evolving Coast*. Scientific American Library, New York.

Dawson, S. and Smith, D. E., 1998. 'Holocene relative sea-level changes on the margin of a glacio-isostatically uplifted area: an example from northern Caithness, Scotland'. *The Holocene*, 7, 59–77.

Deford, F., 1999. 'The Florida Keys: Paradise with attitude'. *National Geographic*, 196(6), 32–53.

Dyer, K. R., 1997. *Estuaries: A Physical Introduction* (2nd Edition). Wiley, Chichester.

El Raey, M., El Din, S. H. S., Khafagy, A. A. and Zed, A. I. A., 1999a. 'Remote sensing of beach erosion/accretion patterns along Damietta-Port Said shoreline, Egypt'. *International Journal of Remote Sensing*, 20, 1087–1106.

El Raey, M., Frihy, O., Nasr, S. M. and Dewidar, K. H., 1999b. 'Vulnerability assessment of sea level rise over Port Said Governorate, Egypt'. *Environmental Monitoring and Assessment*, 56, 113–128.

Environment Agency, 1999. 'Ice victim Beachy slips into the stormy sea'. *Environment Action*, 18, 5.

Feng, J. L. and Zhang, W., 1998. 'The evolution of the modern Luanhe River Delta, north China'. *Geomorphology*, 25, 269–278.

Fera, P., 1993. 'Marine eutrophication along the Brittany coasts – origin and evolution'. *European Water Pollution Control*, 3, 26–32.

French, P. W., 1997. *Coastal and Estuarine Management*. Routledge, London.

Galloway, W. E., 1975. 'Process framework for describing the morphologic and stratigraphic evolution of deltaic depositional systems'. In: M. L. Broussard (ed.) *Deltas: Models for Exploration*. Houston Geological Society, Houston, 87–98.

Galvin, C. J., 1968. 'Breaker-type classification on three laboratory beaches'. *Journal of Geophysical Research*, 73, 3651–3659.

Gehrels, W. R., 1994. 'Determining relative sea-level change from salt-marsh foraminifera and plant zones on the coast of Maine, USA'. *Journal of Coastal Research*, 10, 990–1009.

Gehrels, W. R., Belknap, D. F. and Kelley, J. T., 1996. 'Integrated high-precision analyses of Holocene relative sea-level changes: lessons from the coast of Maine'. *Geological Society of America Bulletin*, 108, 1073–1088.

Gilbert, G. K., 1885. 'The topographic features of lake shores'. *Annual Reports of the US Geological Survey*, 5, 75–123.

Giosan, L., Bokuniewicz, H., Panin, N. and Postolache, I., 1999. 'Longshore sediment transport pattern along the Romanian Danube delta coast'. *Journal of Coastal Research*, 15, 859–871.

González, F. I., 1999. 'Tsunami!' *Scientific American*, May, 44–55.

Goodbred, S. L. and Kuehl, S. A., 1999. 'Holocene and modern sediment budgets for the Ganges–Brahmaputra river system: evidence for highstand dispersal to flood-plain, shelf, and deep-sea depocentres'. *Geology*, 27, 559–562.

Gornitz, V., 1995. 'Sea-level rise: a review of recent past and near-future trends'. *Earth Surface Processes and Landforms*, 20, 7–20.

Gornitz, V., Rosenzweig, C. and Hillel, D., 1997. 'Effects of anthropogenic intervention in the land hydrologic cycle on global sea level rise'. *Global and Planetary Change*, 14, 147–161.

Guilcher, A., Hallégouet, B., Meur, C., Talec, P. and Yoni, C, 1992. 'Exceptional formation of present-day dunes in the Baie d'Audierne, southwestern Brittany, France'. In: R. W. G. Carter (ed) *Coastal Dunes. Proc. 3rd European Dune Congress, Galway, 1992.* (Balkema), 15–23.

Haslett, S. K., Bryant, E. A. and Curr, R. H. F., 2000a. 'Tracing beach sand provenance and transport using foraminifera: preliminary examples from NW Europe and SE Australia'. In: I. Foster (ed.) *Tracers in Geomorphology.* Wiley, Chichester, 437–452.

Haslett, S. K. and Curr, R. H. F., 1998. 'Coastal rock platforms and Quaternary sea-levels in the Baie d'Audierne, Brittany, France'. *Zeitschrift für Geomorphologie N. F.,* 42, 507–515.

Haslett, S. K., Davies, P., Curr, R. H. F., Davies, C. F. C., Kennington, K., King, C. P. and Margetts, A. J., 1998a. 'Evaluating late-Holocene relative sea-level change in the Somerset Levels, southwest Britain'. *The Holocene,* 8, 197–207.

Haslett, S. K., Davies, P. and Strawbridge, F., 1998b. 'Reconstructing Holocene sea-levels in the Severn Estuary and Somerset Levels: the foraminifera connection'. *Archaeology in the Severn Estuary,* 8, 29–40.

Haslett, S. K., Davies, P. and Curr, R. H. F., 2000b. 'Geomorphologic and palaeo-environmental development of Holocene perched coastal dune systems in Brittany, France'. *Geografiska Annaler, Series A,* 82, 79–88.

Healy, T., 1991. 'Coastal erosion and sea level rise'. *Zeitschrift für Geomorphologie N. F., Supplementband,* 81, 15–29.

Hensel, P. F., Day, J. W. and Pont, D., 1999. 'Wetland accretion and soil elevation change in the Rhône River Delta, France: the importance of riverine flooding'. *Journal of Coastal Research,* 15, 668–681.

Hooke, J. M. and Bray, M. J., 1995. 'Coastal groups, littoral cells, policies and plans in the UK'. *Area,* 27, 358–368.

Horsfall, D., 1993. 'Geological controls of coastal morphology'. *Geography Review,* 7(1), 16–22.

Horton, B. P., Edwards, R. J. and Lloyd, J. M., 1999. 'UK intertidal foraminiferal distributions: implications for sea-level studies'. *Marine Micropalaeontology,* 36, 205–223.

Houghton, J. T., 1994. *Global Warming: The Complete Briefing.* Lion Books, Oxford.

Houghton, J. T., Jenkins, G. J. and Ephraums, J. J. (eds), 1990. *Climate Change: The IPCC Scientific Assessment.* Cambridge University Press, Cambridge.

Houghton, J. T., Meira Filho, L. G., Callander, B. A., Harris, N., Kattenberg, A. and Maskell, K. (eds) 1996. *Climate Change 1995: The Science of Climate Change (contribution of WGI to the second assessment report of the Intergovernmental Panel on Climate Change).* Cambridge University Press, Cambridge.

Huybrechts, P., 1990. 'A 3-D model for the Antarctic ice sheet: a sensitivity study on the glacial–interglacial contrast'. *Climate Dynamics,* 5, 79–92.

Inman, D. and Nordstrom, C., 1971. 'On the tectonic and morphological classification of coasts'. *Journal of Geology,* 79, 1–21.

Kay, R. and Alder, J., 1999. *Coastal Planning and Management.* E & FN Spon, London.

Knighton, A. D., Woodroffe, C. D. and Mills, K., 1992. 'The evolution of tidal creek networks, Mary River, Northern Australia'. *Earth Surface Processes and Landforms,* 17, 167–190.

Komar, P. D., 1976. *Beach Processes and Sedimentation*. Prentice-Hall, Englewood Cliffs.

Laborel, J. and Laborel-Deguen, F., 1994. 'Biological indicators of relative sea-level variations and co-seismic displacements in the Mediterranean region'. *Journal of Coastal Research*, 10, 395–415.

Laborel, J., Morhange, C., Lafont, R., Le Campion, J., Laborel-Deguen, F. and Sartoretto, S., 1994. 'Biological evidence of sea-level rise during the last 4,500 years on the rocky coasts of continental southwestern France and Corsica'. *Marine Geology*, 120, 203–223.

Lambeck, K., 1995. 'Late Devensian and Holocene shorelines of the British Isles and North Sea from models of glacio-hydro-isostatic rebound'. *Journal of the Geological Society, London*, 152, 437–448.

Larcombe, P. and Carter, R. M., 1998. 'Sequence architecture during the Holocene transgression: an example from the Great Barrier Reef shelf, Australia'. *Sedimentary Geology*, 117, 97–121.

Long, A. J., Roberts, D. H. and Wright, M. R., 1999. 'Isolation basin stratigraphy and Holocene relative sea-level change on Arveprinsen Ejland, Disko Bugt, West Greenland'. *Journal of Quaternary Science*, 14, 323–345.

Long, A. J. and Shennan, I., 1998. 'Models of rapid relative sea-level change in Washington and Oregon, USA'. *The Holocene*, 8, 129–142.

Lowe, J. J. and Walker, M. J. C., 1997. *Reconstructing Quaternary Environments* (2nd Edition). Longman, Harlow.

McBride, R. A., Byrnes, M. R. and Hiland, M. W., 1995. 'Geomorphic response-type model for barrier coastlines: a regional perspective'. *Marine Geology*, 126, 143–159.

McBride, R. A., Penland, S., Jaffe, B., Williams, S. J., Sallenger, A. H. Jr. and Westphal, K., 1991. *Shoreline Changes of the Isles Derniers Barrier Island Arc, Louisiana, from 1853 to 1989*. MAP I-2186 (1:75,000), Miscellaneous Investigations Series, US Geological Survey.

McCullagh, P., 1978. *Modern Concepts in Geomorphology*. Oxford University Press, Oxford.

Mannion, A. M., 1999. *Natural Environmental Change*. Routledge, London.

Masselink, G., 1999. 'Alongshore variation in beach cusp morphology in a coastal embayment'. *Earth Surface Processes and Landforms*, 24, 335–347.

Masselink, G. and Pattiaratchi, C. B., 1998. 'Morphological evolution of beach cusps and associated swash circulation patterns'. *Marine Geology*, 146, 93–113.

Mathers, S. and Zalasiewicz, J., 1999. 'Holocene sedimentary architecture of the Red River Delta, Vietnam'. *Journal of Coastal Research*, 15, 314–325.

Meier, M. F., 1984. 'Contribution of small glaciers to global sea level'. *Science*, 226, 1418–1421.

Michel, D. and Howa, H. L., 1999. 'Short-term morphodynamic response of a ridge and runnel system on a mesotidal sandy beach'. *Journal of Coastal Research*, 15, 428–437.

NERC, 1991. *United Kingdom Digital Marine Atlas*. Natural Environment Research Council, Swindon.

Newman, W. S. and Fairbridge, R. W., 1986. 'The management of sea level rise'. *Nature*, 320, 319–321.

Nichols, M. M. and Biggs, R. B., 1985. 'Estuaries'. In: R. A. Davis (ed.) *Coastal Sedimentary Environments*. Springer-Verlag, New York, 77–186.

Nield, T., 1999. 'Beachy feet'. *Geoscientist*, 9(3), 12.

Nittrouer, C. A., Kuehl, S. A., De Master, D. J. and Kowsmann, R. O., 1986. 'The deltaic nature of Amazon shelf sedimentation'. *Bulletin of the Geological Society of America*, 97, 444–458.

O'Reagan, P., 1996. 'The use of contempory information technologies for coastal research and management – a review'. *Journal of Coastal Research*, 12, 192–204.

Orford, J. D., Carter, R. W. G., Forbes, D. L. and Taylor, R. B., 1988. 'Overwash occurrence consequent on morphological change following lagoon inlet closure on a coarse clastic barrier'. *Earth Surface Processes and Landforms*, 13, 27–36.

Orton, G. J. and Reading, H. G., 1993. 'Variability of deltaic processes in terms of sediment supply, with particular emphasis on grain size'. *Sedimentology*, 40, 475–512.

Park, C., 1997. *The Environment: Principles and Applications*. Routledge, London.

Perillo, G. M. E. (ed.), 1995. *Geomorphology and Sedimentology of Estuaries*. Elsevier, Amsterdam.

Pethick, J., 1984. *Introduction to Coastal Geomorphology*. Arnold, London.

Pethick, J., 1993. 'Shoreline adjustments and coastal management – physical and biological processes under accelerated sea-level rise'. *Geographical Journal*, 159, 162–168.

Pickering, K. T. and Owen, L. A., 1997. *An Introduction to Global Environmental Issues* (2nd Edition). Routledge, London.

Pirazzoli, P. A., 1996. *Sea level Changes: The Last 20000 Years*. Wiley, Chichester.

Populus, J., Moreau, F., Coquelet, D. and Xavier, J. P., 1995. 'An assessment of environmental sensitivity to marine pollutions – solutions with remote sensing and geographical information systems (GIS)'. *International Journal of Remote Sensing*, 16, 3–15.

Postma, G., 1990. 'Depositional architecture and facies of river and fan deltas: a synthesis'. In: A. Colella and D. B. Prior (eds) *Coarse-grained Deltas*. Special Publication of the International Association of Sedimentologists, 10, 13–27.

Postma, H., 1961. 'Transport and accumulation of suspended matter in the Dutch Wadden Sea'. *Netherlands Journal of Sea Research*, 1, 148–190.

Pritchard, D. W., 1955. 'Estuarine circulation patterns'. *Proceedings of the American Society of Civil Engineers*, 81(717).

Raymo, M. E. and Ruddiman, W. F., 1992. 'Tectonic forcing of late Cenozoic climate'. *Nature*, 359, 117–122.

Reading, H. G. and Collinson, J. D., 1996. 'Clastic coasts'. In: H. G. Reading (ed.) *Sedimentary Environments: Processes, Facies and Stratigraphy* (3rd Edition). Blackwell Science, Oxford, 154–231.

Reed, D. J., 1995. 'The response of coastal marshes to sea-level rise: survival or submergence?' *Earth Surface Processes and Landforms*, 20, 39–48.

Reinson, G. E., 1992. 'Transgressive barrier island and estuarine systems'. In: R. G. Walker and N. P. James (eds) *Facies Models: Response to Sea-level Change*. Geological Association of Canada, Waterloo (Ontario), 179–194.

Saleh, A., Al-Ruwaih, F., Al-Reda, A. and Gunatilaka, A., 1999. 'A reconnaissance study of a clastic coastal sabka in Northern Kuwait, Arabian Gulf'. *Journal of Arid Environments*, 43, 1–19.

Scott, D. B. and Medioli, F. S., 1978. 'Vertical zonations of marsh foraminifera as accurate indicators of former sea-levels'. *Nature*, 272, 528–531.

Scott, D. B. and Medioli, F. S., 1986. 'Foraminifera as sea-level indicators'. In: O. van de Plassche (ed.) *Sea-level Research: A Manual for the Collection and Evaluation of Data*. Geo-Books, Norwich, 435–456.

Shennan, I., Innes, J. B., Long, A. J. and Zong, Y., 1994. 'Late Devensian and Holocene relative sea-level changes at Loch nan Eala, near Arisaig, northwest Scotland'. *Journal of Quaternary Science*, 9, 261–283.

Smith, D. E. and Dawson, A. G., 1990. 'Tsunami waves in the North Sea'. *New Scientist*, 4 August, 46–49.

Stowe, K., 1996. *Exploring Ocean Science* (2nd Edition). Wiley, New York.

Stringer, W. J., Dean, K. G., Guritz, R. M., Garbeil, H. M., Groves, J. E. and Ahlnaes, K., 1992. 'Detection of petroleum spilled from the MV Exxon Valdez'. *International Journal of Remote Sensing*, 13, 799–824.

Stuiver, M. and Reimer, P. J., 1993. 'Extended ^{14}C data base and revised CALIB. 3.0 ^{14}C age calibration program'. *Radiocarbon*, 35, 215–230.

Suhayda, J. N. and Pettigrew, N. R., 1977. 'Observations of wave height and celerity in the surf zone'. *Journal of Geophysical Research*, 82, 1419–1424.

Summerfield, M. A., 1991. *Global Geomorphology*. Longman, Harlow.

Sunamura, T., 1992. *Geomorphology of Rocky Coasts*. Wiley, Chichester.

Swift, D. J. P., 1975. 'Barrier-island genesis: evidence from the central Atlantic shelf, eastern USA'. *Sedimentary Geology*, 14, 1–43.

Thomson, R. E. and Tabata, S., 1987. 'Steric height trends of ocean station PAPA in the northeast Pacific Ocean'. *Marine Geodesy*, 11, 103–113.

Trenhaile, A. S., 1999. 'The width of shore platforms in Britain, Canada, and Japan'. *Journal of Coastal Research*, 15, 355–364.

Viles, H. and Spencer, T., 1995. *Coastal Problems: Geomorphology, Ecology and Society at the Coast*. Arnold, London.

Warrick, R. A. and Oerlemans, J., 1990. 'Sea level rise'. In: J. T. Houghton, G. J. Jenkins and J. J. Ephraums (eds) *Climate Change, The IPCC Assessment*. Cambridge University Press, Cambridge, 257–281.

Warrick, R. A., Le Provost, C., Meier, M. F., Oerlemans, J. and Woodworth, P. L., 1996. 'Changes in sea level'. In: J. T. Houghton, L. G. Meira Filho, B. A. Callander, N. Harris, A. Kattenberg and K. Maskell (eds) *Climate Change 1995: The Science of Climate Change (Contribution of WGI to the Second Assessment Report of the Intergovernmental Panel on Climate Change)*. Cambridge University Press, Cambridge, 359–405.

Werner, B. T. and Fink, T. M., 1993. 'Beach cusps as self-organised patterns'. *Science*, 260, 968–970.

White, K. and El Asmar, H. M., 1999. 'Monitoring changing position of coastlines using Thematic Mapper imagery, an example from the Nile Delta'. *Geomorphology*, 29, 93–105.

Wigley, T. M. L. and Raper, S. C. B., 1993. 'Future changes in global mean temperature and sea level'. In: R. A. Warrick, E. M. Barrow and T. M. L. Wigley (eds) *Climate and Sea Level: Observations, Projections and Implications*. Cambridge University Press, Cambridge, 111–133.

Williams, A. T., 1992. 'The quiet conservators: Heritage Coasts of England and Wales'. *Ocean and Coastal Management*, 17, 151–167.

Williams, A. T. and Caldwell, N. E., 1988. 'Particle size and shape in pebble-beach sedimentation'. *Marine Geology*, 82, 199–215.

Williams, A. T. and Simmons, S. L., 1997. 'Estuarine litter at the river/beach interface in the Bristol Channel, United Kingdom'. *Journal of Coastal Research*, 13, 1159–1165.

Woodroffe, C. D., 1993. 'Geomorphological and climatic setting and the development of mangrove forests'. In: H. Lieth and A. Al Masoom (eds) *Towards the Rational Use of High Salinity Tolerant Plants*. Kluwer Academic Publishers, Dordrecht, 13–20.

Woodroffe, C. D., 1995. 'Response of tide-dominated mangrove shorelines in Northern Australia to anticipated sea-level rise'. *Earth Surface Processes and Landforms*, 20, 65–85.

Young, R. W., White, K. L. and Price, D. M., 1996. 'Fluvial deposition on the Shoalhaven Deltaic Plain, southern New South Wales'. *Australian Geographer*, 27, 215–234.

Zong, Y., 1998. 'Diatom records and sedimentary responses to sea-level change during the last 8,000 years in Roudsea Wood, northwest England'. *The Holocene*, 8, 219–228.

Index